7⁹⁸
N
USA++

Einstein and Weizmann in New York.
Courtesy Central Zionist Archives, Jerusalem.

Albert Meets America

✦

Albert Meets America

✦

How Journalists Treated Genius

during Einstein's 1921 Travels

Edited by

JÓZSEF ILLY

The Johns Hopkins University Press
Baltimore

© 2006 The Johns Hopkins University Press
All rights reserved. Published 2006
Printed in the United States of America on acid-free paper
2 4 6 8 9 7 5 3 1

The Johns Hopkins University Press
2715 North Charles Street
Baltimore, Maryland 21218-4363
www.press.jhu.edu

Library of Congress Cataloging-in-Publication Data

Albert meets America : How journalists treated genius during Einstein's 1921 travels /
edited by József Illy.
p.   cm.
Includes bibliographical references and index.
ISBN 0-8018-8457-8 (acid-free paper)
1. Einstein, Albert, 1879–1955—Public opinion. 2. Einstein, Albert, 1879–1955—Travel—
United States. 3. United States—Description and travel. I. Illy, József, 1933–
QC16.E5E44 2006
530.092—dc22
2006005266

A catalog record for this book is available from the British Library.

To Marci
with I+Sz.

✦

Contents

✦

Foreword

✦

When chided for agreeing to make his first trip to the United States as part of a Zionist delegation, rather than as a representative of German science, Einstein replied to his colleague Fritz Haber, the Nobel Prize winner in chemistry:

> Despite my expressed internationalist orientation, I nevertheless always consider it my duty to intervene on behalf of my persecuted and morally oppressed tribal colleagues, as much as is in my power. I therefore gladly agreed [to the request to travel to the United States with a Zionist delegation], without debating for more than five minutes, although I had just turned down all the American universities. This therefore is rather an act of loyalty than one of disloyalty. The prospect of erecting a Jewish university in particular fills me with special joy, since I have recently seen numerous examples of the perfidious and loveless manner in which one treats splendid young Jews and seeks to sever their educational possibilities. I could also list other events of the past year that would have to drive a self-respecting Jew to take Jewish solidarity more seriously than seemed proper and natural in earlier days.

A few days later Einstein reiterated in a letter to his close friend Heinrich Zangger: "On Sunday it's off to America. Not only to speak at universities, which will happen as well, but for the founding of the Jewish university in Jerusalem. I feel the keen need to do something for this cause."

Thus, Albert Einstein, who had received numerous invitations to lecture in America over the preceding few years, traveled from Berlin via the Netherlands in the spring of 1921 with a full agenda and schedule: he would spend most of his time accompanying Chaim Weizmann on a tour of East Coast cities, making numerous, brief appearances at large gatherings intended to arouse enthusiasm for the Jewish colony in Palestine and its planned cultural and educational institutions. In addition, Einstein was to deliver a series of lectures at Princeton University on his contributions to modern physics.

Einstein had been propelled to international fame in the fall of 1919, at age forty, with the confirmation of his general theory of relativity by two teams of British astronomers who had observed the 1919 solar eclipse and produced evidence confirming Einstein's prediction of light bending in the vicinity of massive objects. He had been celebrated as the greatest physicist of the twentieth century on the front pages of all major American and European newspapers.

As one of only a handful of German scientists who had publicly criticized Germany's war aims, Einstein could have expected to travel to former Allied countries without fearing overt anti-German sentiment. Since the end of World War I, however, he had not been officially invited except by formerly neutral or Central power European countries, such as Holland, Switzerland, and Austria. Although the British Royal Astronomical Society had considered awarding him their Gold Medal in 1920, the nomination had ultimately failed to gain the majority needed, and Einstein's planned trip to England to receive this honor had been canceled.

During 1920, Einstein received several invitations to teach or lecture in America. He toyed with the possibility of an extended lecture tour and exchanged letters with representatives of Columbia University, the University of Wisconsin, Harvard University, Princeton University, and the National Academy of Sciences in Washington, D.C. By the end of the year, Einstein was even studying English pronunciation and was expecting to travel soon to the United States. But by early 1921, it had become clear that, in the climate of postwar economic hardships, not even American institutions of higher learning could come up with the ambitious sum of fifteen thousand dollars that Einstein was hoping to be paid for his U.S. lectures.

Distinguished European scientists had begun traveling to America in the previous two decades, received enthusiastically by an academic community eager to adopt the practice of research-oriented scientific education that had taken hold, primarily at German universities, since the late nineteenth century. Physics, and theoretical physics in particular, had experienced extraordinary growth in Germany since Hermann von Helmholtz, Felix Klein, Max Planck, and other distinguished physicists had started the intellectual revolution leading eventually to quantum physics and relativity, fields in which Einstein had made his most remarkable and lasting contributions between 1905 and 1915. It had been customary for American students to travel to the great chemical and physical laboratories of Berlin or Leipzig for postgraduate work and then to return and recast American science departments at MIT or Caltech on the model of German higher education, emphasizing independent research as an integral component of scientific education.

At the 1904 World's Fair in Saint Louis, a stellar gathering of European scientists had delivered comprehensive lectures on the state of science at the turn of the century, among them Max Weber, Jacques Loeb, and Wilhelm Ostwald. A year later, the eminent Austrian physicist Ludwig Boltzmann returned to the United States and ventured as far as Berkeley and the newly founded Stanford University, where he lectured in English and was frequently entertained by munificent donors, among them Mrs. Hearst and Mrs. Leland Stanford. In an essay published that same year as "Trip of a German Professor to Eldorado," Boltzmann gave a vivid and entertaining account of his summer sojourn in California, of American railroads, restaurants, prohibition, and the ubiquitous presence of a Euclid Avenue in every city. Boltzmann was most impressed by the giant telescope of the Lick Observatory and the entrepreneurial spirit of American scientists and their supporters and less impressed by the students in his class, whom he found only modestly prepared to deal with integrals and differentials.

Over the ensuing years, many other European physicists accepted invitations to lecture at major American universities engaged in the new progressive movement of social, economic, and increasingly vibrant educational reform. With these reforms came growth in federal support of science. In the early 1900s, through the efforts of the newly founded American Physical Society and scientists at Princeton, Columbia, Wisconsin, Harvard, Cornell, and other institutions, physics departments expanded their laboratories, faculty, and student enrollment, and "as though to symbolize the vitality of American physics, in 1907 Albert Michelson won his Nobel Prize in physics."[*]

By 1914, the new atomic physics discovered in England and on the Continent was being taught at Princeton, at Chicago (where Robert Millikan carried out his famous oil-drop experiments), at Harvard, and elsewhere. Support for scientific research was growing through private funding agencies, such as the Carnegie Institution, and industrial research laboratories at companies such as AT&T, DuPont, Westinghouse, and General Electric

---

[*]For a detailed account of the development of American physics in the years before and after Einstein's visit, see especially Daniel J. Kevles, *The Physicists: The History of a Scientific Community in Modern America* (Cambridge: Harvard University Press, 1995), 79.

were increasingly attracting talented young physicists. But it was the outbreak of World War I that propelled American science onto its path of ascendancy and into the preeminent position it was to occupy for the rest of the twentieth century.

The war caused a major rupture in international scientific relations, with wide-ranging repercussions for the further development of science. With the recruitment by all belligerent countries of engineers, chemists, and physicists in the service of the most devastating and technologically advanced conflagration in history, the reestablishment of scholarly exchanges among former enemy nations after 1919 was a slow, cautious process, fraught with resentment for many. Marie Curie, the two-time Nobel Prize winner in chemistry and physics (who had introduced x-ray machines on ambulances on the French front), was received triumphantly on May 20, 1921, at the White House, where she was presented with a gift of one gram of radium by President Harding. Einstein, however, was not invited but had to content himself with meeting Harding as part of a delegation of scientists. Even two years after the end of the war, his German origin and residence caused many officials to be cautious. Nevertheless, as the articles so conscientiously collected and humorously annotated by Dr. Illy in the present volume indicate, many representatives of local and state governments and of academic establishments along the East Coast sincerely welcomed Einstein.

During his U.S. travels, Einstein spent considerable time in New York City and also visited New Brunswick, Washington, D.C., Chicago, Princeton, Boston, New Haven, and Cleveland. He lectured at the College of the City of New York, Columbia, Princeton, the American Academy of Arts and Sciences, and the Zionist Society of Engineering and Agriculture. From the more than 160 newspaper reports contained in this comprehensive collection, however, one senses that he tried hardest to explain to reporters and the general public, invariably through an interpreter, the essence of his theory of relativity, its significance, and its usefulness.

This collection highlights the creation of the public image of Einstein, and of relativity, at the hands of the press, the public, and of Einstein himself. His physical and emotional attributes, his pleasant, jocular, and at times exasperated demeanor, his professorial attire and peculiarities, his eating habits and opinions on America were all dissected, reported, and repeated. Einstein's wife, Elsa, is also fully present in many articles. She apparently served as Einstein's interpreter on numerous occasions and also made her own mark with the press and the various dignitaries they encountered. Her English skills were evidently much better than Einstein's, and she relished entertaining, recounting anecdotes, and replying to many questions regarding Einstein's work habits in Berlin, his tastes and preferences, his likes and dislikes. She comes across as an energetic, determined woman, the worldly companion to the maladroit savant.

The two months that Einstein spent in the United States cemented his status as the scientific icon of the twentieth century. Einstein later embarked on a few similar long journeys, to Japan and the Far East and to South America. He returned to America a decade later as a visiting scientist at Caltech for three successive winters, starting in January 1931. In early 1933, while Einstein was in Pasadena, Hitler's Nazi regime made Einstein's return to Germany impossible. His fifth and last transatlantic journey brought Einstein to New York in the fall of 1933, when he took up his last academic position as a member of the newly founded Institute for Advanced Study in Princeton, where he spent the last two decades of his life.

The United States had become his third home country, a country that he supported wholeheartedly during World War II but whose misdeeds during the McCarthy era he criticized equally vigorously. During the interwar years he had served the cause of international reconciliation and pacifism and had thrown the weight of his scientific reputation behind the cause of oppressed and disenfranchised Eastern European Jews. He joyously welcomed the founding of the State of Israel in 1948 but continued to advocate tirelessly on behalf of peaceful coexistence and cooperation between Jews and Arabs. After 1945 the press and the world saw Einstein as the "father of the atomic bomb," on which he in fact never worked. He continued to feel a weight of responsibility for the devastations of Hiroshima and Nagasaki and devoted the last decade of his life to promoting the civilian control of nuclear energy and international nuclear disarmament.

Diana K. Buchwald
General Editor and Director
The Einstein Papers Project

Preface

✦

Sing of Einstein's
Yiddishe peachtrees, sing of
Sleep among the cherryblossoms.
Sing of wise newspapers
That quote the great mathematician:
A little touch of
Einstein in the night—

—William Carlos Williams,
"St. Francis Einstein of the Daffodils" (First Version), 1921

When Einstein joined Chaim Weizmann's tour of the United States in 1921 to gain support among American Jewry for the Zionist cause, his role was to raise funds for the establishment of Hebrew University. Although news coverage of the trip frequently focused on Weizmann's fierce disputes with the leaders of American Zionism, I concentrate on Einstein for three reasons.

First, in spite of Weizmann's leading role in the organization and aims of the trip, the central figure in the American and English-language Zionist press reports was Einstein, who just two years previously had gained worldwide celebrity because of a successful test of his general theory of relativity. As one journalist put it, "There is reason to believe that Prof. Einstein was induced to accompany the mission in the hope that his presence would act as a 'tail to the kite.' And now, lo and behold, contrary to all calculations (except, perhaps, those based on the theory of relativity), the tail has become the whole kite."

Second, Einstein did not participate in the fierce debate between the Weizmann-led World Zionist Organization and the leaders of the Zionist Organization of America (the "Brandeis-group"), although Einstein followed the debate and supported the views of Weizmann. The two salient points of the debate were (1) whether to build up the Jewish national home in Palestine with private (Brandeis) or public funds (Weizmann's Keren Hayesod) and (2) whether the Zionist organization should have a federal system with independent member organizations (Brandeis) or the Zionist Congress should be a parliament exerting central control (Weizmann). (You see the parallel with the federalist and unionist viewpoints in American history!) Weizmann won the debate in Cleveland with the help of the majority of American Jews, the "Easterners" (i.e., those with roots in Eastern Europe), against the official leaders of American Jewry, who were mostly "Westerners" (i.e., of ancestry from Germany).

A third reason for concentrating on Einstein is somewhat subjective. Working for the Einstein Papers Project, I have learned that one should be careful with newspaper accounts. As historical scholars, we collect each scrap of newspaper with Einstein's name in it, but when forming scholarly conclusions, we turn to them only as a last resort. As a result, we now have a sizable collection of clippings, with interesting interviews, events, and commentaries. We keep them in big, black steel cabinets that protect them from floods, fires, earthquakes, and the eyes of readers, I fear. As a conservation-minded person, I cannot tolerate this waste of intellectual treasure and interesting reading.

A few words about the news reports presented herein. I had to use microfilm copies of the originals, often on the edge of legibility; hence transcription was the only feasible option for their presentation.

Names are another problem. The reporters recorded the names by following their ears. Weitzmann, Weisemann, or Weizmann? These variants cause no trouble because their contexts show they all stand for the name of a well-known person, Chaim Weizmann. But what about Bernstein/Burstein, Hurwitz/Hurevitz, Levin/Lewine, Timen/Teaman? I have left the spelling variants untouched and apologize if any find their ancestors' names misspelled.

You might well wonder whether Einstein really was the kind of Zionist that the journalists portrayed. For a recent discussion of this fundamental issue, please see the seventh volume of *The Collected Papers of Albert Einstein.*

I enjoyed gathering these materials. I hope you will enjoy reading them, even if not everything that was reported on Einstein is true. Indeed, you may find pleasure in unraveling the roots of the errors—as did I.

I express my sincere thanks to Robert Schulmann and Diana Kormos-Buchwald, for encouraging the whole venture; to Jane Dietrich, Rudy Hirschmann, and Rosy Meiron for translating my English into theirs; to Ze'ev Rosenkranz, for help in finding my way in Jewish matters; and to Daniel Kennefick, for his remarks on Irish matters.

I am grateful to Dr. Roni Grosz, curator of the Albert Einstein Archives, for permission to publish the manuscript "Einstein on the Art of Interviewing."

In collecting the sources and illustrations, I received indispensable help from Maurita Baldock, New York University; Jeff Bridgers, Library of Congress; Carol Butler, Brown Brothers, Sterling, Pennsylvania; J. Frank Cook, University of Wisconsin; Erin M. Cosyn, Johns Hopkins University Press; Rob Cox, American Philosophical Society; Richard Dreiser, Yerkes Observatory; Nancy Finlay, Connecticut Historical Society, Hartford; Walter G. Heverly, University of Pittsburgh; Stella Hsu, National Institute of Standards; Barbara Kern, University of Chicago; Reuven Koffler, Central Zionist Archive, Jerusalem; Liezl Lao, Corbis Los Angeles; Heather Lindsay, Emilio Segrè Visual Archives, American Institute of Physics; Norbert Ludwig, Bildarchiv Preußischer Kulturbesitz, Berlin; Karen Murphy, New York University; Clifford L. Muse Jr., Howard University; Bill O'Hanlon, Stanford University Libraries; Shadye Peyvan, Millikan Library, California Institute of Technology; Linda J. Porter, Pittsfield, Massachusetts; Roberto Trujillo, Stanford University Libraries; and Sydney C. Van Nort, City College of the City University of New York.

Special thanks to Martin Schneider, the copyeditor of the book, for his knowledgeable remarks and advice.

I am deeply grateful to Alice Calaprice for her invaluable help, which made the publication of the book possible.

Albert Meets America

✦

1

Antecedents

✦

The first official step taken by the World Zionist Organization to request Einstein's help in fostering the cause of Hebrew University in Jerusalem was a letter written by Hugo Bergmann on October 22, 1919. Bergmann asked Einstein to join a fundraising mission to the United States for the Palestine Foundation Fund (*Keren Hayesod*).[1]

Einstein's earlier informal talks with German Zionists and his public statements in the press were a strong indication that an invitation in the university matter would not fall on deaf ears. Indeed, four days after Chaim Weizmann, president of the World Zionist Organization, had sent a telegram to Kurt Blumenfeld, a leading German Zionist, asking him to invite Einstein on his behalf to join the mission,[2] Blumenfeld sent the following reply: "Einstein prepared to join you for America letter follows."[3]

2

To Visit America

✦

February 21

Prof. Einstein to Visit America

Berlin, Feb. 21—Prof. Albert Einstein, distinguished scientist, whose theory of relativity has evoked tremendous interest throughout the world, will come to America next month with the Zionist delegation. It is announced that he will appeal to Jews of America for support of the Hebrew University to be erected on the Mount of Olives.

The Jewish Independent, February 25, p. 1.

March 4

Dr. Weizmann and Prof. Einstein Coming to America Soon

Dr. and Mrs. Chaim Weizmann and Prof. and Mrs. Albert Einstein will sail for America on March 24th, according to a cable received by the Zionist office. They will be accompanied by a group of European Zionists.

Prof. Einstein will return to Holland later in April, where he has promised to deliver a series of lectures in Dutch universities. He is extremely interested in the Hebrew University, and has offered his services to that university.

The New Palestine, March 4, p. 1.

March 25

Weizmann and Einstein to Arrive April 3

The President of the World Zionist Organization, Dr. Chaim Weizmann, will arrive in the United States, accompanied by a number of prominent European Zionists, including Professor Einstein, the world-renowned scientist, M. M. Ussichkin, Naiditch and Hillen Zlatapolsky on April 3. This will be Dr. Weizmann's first visit to the United States.

The American Hebrew, March 25, p. 528.

March 26

How Einstein, Thinking in Terms of the Universe, Lives from Day to Day

Prof. Albert Einstein, the distinguished German physicist, whose theory of relativity has made his name a familiar one throughout the world is to sail for the United States on March 28. He comes to lecture on his theory.

By Elias Tobenkin

Too much nationalism—that is the disease from which mankind is suffering to-day. To become normal again the world must return to its pre-war internationalism. Such is the diagnosis of Europe's and the world's ills by Germany's foremost physical scientist, Albert Einstein, discoverer of the Einstein theory of relativity.

I had come to Prof. Einstein to hear what he had to say about the "plight of German science," a subject which just then occupied much space in the newspapers of Berlin. Prof. Einstein, however, spoke not of science but of humanity.

"Of course," he said, "science is suffering from the terrible effects of the war, but it is humanity that should be given first consideration. Humanity is suffering in Germany, everywhere in Eastern Europe as it has not suffered in centuries.

"Humanity," he continued, "is suffering from too much and too narrow a conception of nationalism. The present wave of nationalism, which at the slightest provocation or without provocation passes over into chauvinism, is a sickness.

"The internationalism that existed before the war, before 1914, the internationalism of culture, the cosmopolitanism of commerce and industry, the broad tolerance of ideas, this internationalism was essentially right. There will be no peace on earth, the wounds inflicted by the war will not heal, until this internationalism is restored."

"Does this imply that you oppose the formation of small nations?"

"Not in the least," he replied. "Internationalism as I conceive it implies a rational relationship between countries, a sane union and understanding between nations, mutual coöperation, mutual advancement without interference with a country's customs or inner life."

"And how would you proceed to bring back this internationalism that existed prior to 1914?"

Science and Internationalism

"Here," he said, "is where science, scientists, and especially the scientists of America, can be of great service to humanity. Scientists, and the scientists of America in the first place, must be pioneers in this work of restoring internationalism.

"America is already in advance of all other nations in the matter of internationalism. It has what might be called an international 'psyche.' The extent of America's leaning to internationalism was shown by the initial success of Wilson's ideas of internationalism, the popular acclaim they met with the American people.

"That Wilson failed to carry out his ideas is beside the point. The enthusiasm with which the preaching of these ideas by Wilson was received shows the state of mind of the American public. It shows it to be internationally inclined. American scientists should be among the first to attempt to develop these ideas of internationalism and to help carry them forward. For the world and that means America also, needs a return to international friendship. The work of peace cannot go forward in your own country, in any country, so long as your Government or any Government is uneasy about its international relations. Suspicion and bitterness are not a good soil for progress. They should vanish. The intellectuals should be among the first to cast them off."

There are two men in Germany today who are traditionally inaccessible to newspaper men. One is the financier Hugo Stinnes. The other is Einstein.

Einstein has been greatly abused by a section of German press and he therefore shuns publicity. He lives in a quiet section of Berlin on the top floor of a fairly up-to-date apartment house. His study consists of a reception room, or rather a conference room, and of his private workroom. The walls of the conference room are lined with books of a general character. The large number of English books is especially noticeable. There is an edition *de luxe* of Dickens in English, and a costly Shakespeare edition in German. Alongside of Shakespeare stands Goethe in a similarly luxurious edition. Einstein is an admirer of both Goethe and Schiller, and has the busts of the two poets prominently displayed.

Adjoining the conference room is a large music room. When he is not in his study, Mrs. Einstein told me, her husband is in the music room. Music and cigars are the scientist's only relaxations. The number of cigars he smokes is controlled by Mrs. Einstein for his health's sake, but there is no control over the amount of time he chooses to spend at the piano or with his violin, for he plays both instruments well.

His workroom is exceedingly simple. There is a telescope in it. The windows give an exceptionally good view of the sky. There are also a number of globes and various metal representations of the solar system. There are two engravings of Newton on the walls. They are the only pictures in the room. The table he works at is simple and rather small. There is a small typewriter, which is used by his secretary. Einstein has a large correspondence, receiving on an average sixty letters a day.

His Bursts of Concentration

He was pacing up and down the room as I entered his study. He was dressed in a pair of worn-out trousers and a sweater coat. If he had a collar on, the collar was very unobtrusive, for I cannot recall having seen it. He was at work. His hair was dishevelled and his eye had a roving look. His wife told me that when the professor is seized by a problem the fact becomes known to her by this peculiar wandering look which comes into his eyes and by his feverish pacing up and down the room. At such times, she said, the professor is never disturbed. His food is brought to him in his workroom. Sometimes this mode of living lasts for three or four days at a time. It is when the professor rejoins his family at the table that they know that his period of intense concentration and abstraction is over.

After such a period of concentration, Einstein often rests himself by reading fiction. He is fond of reading Dostoyevsky. He walks a lot through the parks and in the summer often goes out with his family in the fields. But he is never asked by his wife or children to go for a walk. It is he who has to do the asking and when he asks them for a walk they know that his mind is relieved of work. His hours of work are indefinite. He sometimes struggles through a whole night with a problem and goes to bed only late in the morning.

Dr. Einstein asked whether he could not see a copy of my interview with him before it was printed. I told him that I would not write the interview until after my return to America.

"In that event," he said, "when you write it, be sure not to omit to state that I am a convinced pacifist, that I believe that the world has had enough of war. Some sort of an international agreement must be reached among nations preventing the recurrence of another war, as another war will ruin our civilization completely. Continental civilization, European civilization, has been badly damaged and set back by this war, but the loss is not irreparable. Another war may prove fatal to Europe."

The New York Evening Post, March 26.

> The article was "advertised" in the Graphic Section, part 4, p. 1 of the *New York Evening Post*.
>
> The famous fourteen points of Woodrow Wilson on the self-determination of small nationalities and on a League of Nations were accepted with enthusiasm by the American public, but in 1918 he failed in convincing Congress to accept the Peace Treaty of Versailles.
>
> Einstein's being "abused by a section of German press" is a hint at the attacks against him and his theory by a group of German physicists and journalists in 1920. He and those on his side were accused of making propaganda for his theories and exerting "mass suggestion."
>
> Tobenkin also mentions Einstein's secretary. At the time Einstein was the director of the Institute of Physics of the Kaiser Wilhelm Society. But one should not think of a big institution with laboratories and an extensive array of instruments: Einstein's lab was actually the den in his own apartment, and the secretary was his elder step-daughter, Ilse Einstein.

The *Rotterdam* on the Einsteins' card to Paul Ehrenfest, March 24, 1920:
"Dear Ehrenfest! It's so fine in the land of milk and honey where our only duty is to find the proper table where we are supposed to eat. Today, however, Zion appears with all its seriousness. Now we enter the Big Ditch. See you in Leyden. With best regards to you all. Your Einstein." (I have not translated Elsa Einstein's note.)[4]
Courtesy Albert Einstein Archives 73-256.

March 30

Prof. Einstein Will Arrive Here April 2

Famous Exponent of Theory of Relativity Coming in Interest of Zionism.

Professor Albert Einstein, exponent of the theory of relativity, will arrive in this city April 2 on the steamship Rotterdam. He will be accompanied by Dr. Chaim Weizmann, formerly professor of chemistry in the University of Manchester and president of the International Zionist Organization.

The two are coming to this country in the interest of Zionism. It was announced yesterday by the Zionist Organization of America.

Mayor Hylan yesterday appointed a representative committee to greet the party. On the committee named are Benjamin Schlesinger, George Gordon Battle and Nathan Strauss. John Dewey, now in China, was also appointed to the welcoming committee by the Mayor.

A mass meeting and reception have been arranged for Sunday afternoon, April 10, at the Metropolitan Opera House.

The New York Call, March 30, p. 2.

April 1

Plan Celebrations in Honor of Visiting Zionist Leaders

Dr. Weizmann, Prof. Einstein and Other Zionist Heads to Reach United States Sunday— Meeting to Be Held Here Sunday—Clevelanders Arrange Gathering for Margolies Memorial Hall—Reception in Honor of Guests to Be Held April 6 and 10 in New York—Mayor Hylan Names Committee

With the expected arrival in the United States Sunday of Dr. Chaim Weizmann, president of the World Zionist Organization, Dr. Albert Einstein, renowned scientist and discoverer of the principle of relativity, M. M. Ussichkin and Dr. Mossensohn, Zionists in many cities throughout the land are planning celebrations in honor of the visit of the distinguished Zionist leaders. This is the first visit of Dr. Weizmann. The party is reported to have left England March 24.

In honor of their arrival a meeting will take place here [in Cleveland] Sunday afternoon at 2 o'clock at the Margolies Hall on E. 55th street. Joseph Barondess is expected to be one of the speakers and addresses will also be made by Isaac Carmel, Rabbi Benjamin and Rabbi Schussheim.

It is expected that the Zionist leaders will visit Cleveland during their stay in the United States and at a conference held last Saturday evening a committee was named to get in touch with the Cleveland Zionist District heads and arrange for a joint council representing all organizations, that would plan a reception for the visitors.

Clevelanders are also planning to attend a celebration in honor of the visitors that is to take place April 6 at the 69th Regiment Armory. Another public reception will take place Sunday, April 10, at the Metropolitan Opera House under the auspices of the Zionist Organization. Mayor Hylan has appointed a citizens' committee to participate in the reception. Nathan Straus and Judge Julian W. Mack will be prominent among those receiving the guests.

Prof. Einstein's visit is in the interest of the Hebrew University to be established in Jerusalem. Dr. Einstein was born in Switzerland in 1879 and has held professorships at Zurich, Prague and Berlin. His work in connection with the subject of relativity brought him world wide fame as a scientist.

Among the New Yorkers who initiated the movement for a popular reception to Professors Weizmann and Einstein and the other members of the delegation are Judge Gustav Hartman, grand master of the I.O.B.A., who is chairman of the committee, Reubin Branin, Louis Lipsky, Rev. Masliansky, Joseph Barondess, Jacob Ish-Kishor, Rabbi Margulies, Dr. Elias Solomon, David Pinsky, Baruch Zuckerman, Professor M. M. Kaplan, Morris Rothenberg, Bernard A. Rosenblatt, Bernard G. Richards, Judge Jacob S. Strahl, Ezekiel Rabinowitz, Yehoash, Sholom Asch.

The Jewish Independent, April 1, p. 1.

Einstein was born in Germany, not in Switzerland. Reconciling his Swiss citizenship with his prestigious position in Berlin proved to be a hard nut for journalists to crack.

The abbreviation "I.O.B.A." stands for "Independent Order of B'rith Abraham" (Abraham's Covenant), a fraternal society founded by American Jews of Eastern European descent (i.e., "Easterners"). See its "Western" counterpart, the Independent Order of B'nai B'rith (Sons of the Covenant) on p. 257.

What happened aboard the *Rotterdam*? The journey was the first opportunity for Weizmann to meet Einstein in person.

"I never met Professor Einstein before this voyage," said Professor Weizmann, who is a great admirer of his fellow-scientist. "He has a singularly sweet and lovable nature, and is exceedingly simple in his habits of life. I have talked with him many times about his work, and he is glad to speak of it when he can find some one who is interested and at least partly capable of understanding it. I do not entirely, for when I get beyond the atom I am lost.

"When he was called 'a poet in science' the definition was a good one. He seems more an intuitive physicist, however. He is not an experimental physicist, and although he is able to detect fallacies in the conceptions of physical science, he must turn his general outlines of theory over to some one else to work out. That would be readily understandable to a man of science. He first became interested in mathematics when he was 14 years old, and his work is his life. He spends most of his time reading and thinking when he is not playing his violin."

The New York Times, April 3, p. 13.

Einstein played the violin even aboard the ship.

On the ship, when a concert was held Dr. Einstein played selections from Mozart, of whose work he is particularly fond, on the violin. Brahms is another of his favorites.

The New York Times, April 3, p. 13.

But the journey was also the first opportunity for Einstein to meet the very attractive and intelligent Mrs. Weizmann. Einstein could not resist her charms. Vera Weizmann turned to Elsa Einstein, asking whether she was embarrassed by Einstein's flirting with her. Elsa's answer was soothing for Vera and surprising for us: "Intellectual women did not attract him. Out of pity he was attracted to women who did manual work."[5]

The *New Palestine*, the official weekly of the Zionist Organization of America, announced the event and introduced the members of the mission.

Over 5,000 people are expected to attend the mass-meeting of welcome which is being given on April 10th to Dr. Chaim Weizmann, president of the World Zionist Organization, and his associates, who are expected to arrive in this country on the Rotterdam, on Sunday, April 3rd. Accompanying Dr. Weizmann are Prof. Albert Einstein, famous for his theory of relativity, who is coming here in the interest of the Hebrew University, M. M. Ussischkin, head of the Zionist Commission in Palestine; Dr. Ben Zion Mossensohn, principal of the Jaffa Gymnasium, Gerson Agronsky, head of the Zionist Commission Press Bureau, Solomon Ginsburg, son of the famous Jewish philosopher Asher Ginzburg (Achad Ha'am), and L. Stein of the London Zionist office.

Hon. Nathan Straus is honorary chairman, and Judge Julian W. Mack is chairman of the committee on arrangements which includes the leaders of every phase of American Jewish life.

A cablegram of greeting to Dr. Weizmann and the Zionist delegation expressing the wish that their negotiations with the American Zionists meet with success, was received by the New York office from the Central Zionist Committee of Warsaw, and signed by Messrs. Klumel, Greenbaum, and Podlishewski. It read:

"Greet Zionist Delegation with Weizmann at head. With success their work. East European Jewry and tens of thousands all over the world are ready to work in Palestine. We await the speediest beginning of the great colonization activity, and hope for the best results of your work."

The New Palestine, April 1, p. 1.

In the following article, I omit the part that introduces Weizmann and Ussischkin.

About the Life and Zionist Activities of Weizmann, Ussischkin and Einstein

. . . Prof. Albert Einstein, the son of a German Jewish family, was born in 1879 in Ulm, Wurtemberg. He spent his schooldays in Munich where he attended a gymnasium. From 1896 to 1900 he studied mathematics and physics at the Technical High School in Zurich, Switzerland. Originally he intended to become a school teacher, but having meanwhile become naturalized, he obtained a post as Engineer in the Swiss Patent Office. The main ideas involved in the most important of Einstein's theories date back to this period.

In 1909 he became "extraordinarius" Professor at the University of Zurich. Later he was called to Prague, Bohemia, to become Professor Ordinarius. In 1913–1914 he accepted a similar chair in Zurich Polytechnicum, when he received an invitation to the Prussian Academy of Science, Berlin. It was in Berlin where he completed his work of "General Theory of Relativity" (1915–1917). Professor Einstein also lectured on various special branches of physics at the University of Berlin and was Director of the Institute for Physical Research of the Kaiser Wilhelm Gesselschaft. He has pledged his services to the Hebrew University in Jerusalem, and it is in the interest of that institution that he is coming here.

The New Palestine, April 1, p. 2.

3

Prof. Einstein Here

✦

Einstein Due Today; Leaders Await Him

Mayor's Committee to Meet Noted Scientist on His Arrival at Quarantine—Big Reception in Hoboken—City's Official Welcome to Take Place on Tuesday and Zionist Greeting on April 10

When Professor Albert Einstein, whose theories have evoked world-wide discussion, arrives today with Professor Chaim Weizmann, chemist and President of the World Zionist Congress, and M. M. Ussischkin, member of the International Zionist Committee, on the Rotterdam of the Holland-America Line, they will be greeted at quarantine by a committee appointed by Mayor Hylan. Jewish Legionnaries who fought in Palestine will march to Pier 2, Hoboken, where the Rotterdam is to dock, to take part in a reception in which it is expected that several thousand persons will participate.

The committee has requested that all Jewish sections of the city be decorated in honor of the visitors. An official meeting of welcome will be held on the steps of City Hall on Tuesday. Among those who will speak are Mayor Hylan, Former Assistant Secretary of State Frank L. Polk, George W. Wickersham and Magistrate Bernard A. Rosenblatt. The American Zionist organization will tender a reception to Professor Einstein, Professor Weizmann and Mr. Ussischkin in the Metropolitan Opera House on Sunday evening, April 10.

Nathan Straus is Honorary Chairman and Magistrate Rosenblatt Chairman of the committee appointed by the Mayor. This committee will go down the bay this morning to greet the visitors and bring them to the

Hoboken piers. Among those on this committee are Arthur Brisbane, Chancellor Elmer Elsworth Brown of New York University, Judge Benjamin Cardoza, ex-Ambassador Abraham I. Elkus, James A. Foley, President F. H. LaGuardia of the Board of Aldermen, Supreme Court Justice Samuel Greenbaum, William D. Guthrie, Mrs. William R. Hearst, ex-Governor Alfred E. Smith, Samuel Koenig, Judge Otto A. Rosalsky, Dr. Bernard Flexner, Benjamin Schlesinger, Oscar S. Straus, Herman Bernstein, George Gordon Battle, Marcus Loew, Adolph Lewisohn, Senator Nathan Straus Jr., and Colonel Robert Greer Monroe. Judge Gustav Hartmann is Chairman of the Provisional Committee.

The New York Times, April 2, p. 11.

Prof. Einstein Here, Explains Relativity

"Poet in Science" Says It Is a Theory of Space and Time, But It Baffles Reporters—Seeks Aid for Palestine—Thousands Wait Four Hours to Welcome Theorist and His Party to America

A man in a faded gray raincoat and a flopping black felt hat that nearly concealed the gray hair that straggled over his ears stood on the boat deck of the steamship Rotterdam yesterday, timidly facing a battery of cameramen. In one hand he clutched a shiny briar pipe and with the other clung to a precious violin. He looked like an artist—a musician. He was.

But underneath his shaggy locks was a scientific mind whose deductions have staggered the ablest intellects of Europe. One of his traveling companions described him as an "intuitive physicist" whose speculative imagination is so vast that it senses great natural laws long before the reasoning faculty grasps and defines them.

The man was Dr. Albert Einstein, propounder of the much-debated theory of relativity that has given the world a new conception of space, and time and the size of the universe.

Dr. Einstein comes to this country as one of a group of prominent Jews, who are advocating the Zionist movement and hope to get financial aid and encouragement for the rebuilding of Palestine and the founding of a Jewish university. He is of medium height, with strong built shoulders, but an air of fragility and self-effacement. Under a high, broad forehead are large and luminous eyes, almost childlike in their simplicity and unworldliness.

The Einsteins on board the *Rotterdam* on arrival in New York.
Library of Congress, courtesy AIP Emilio Segrè Visual Archives.

Thousands Welcome Him

With him as fellow-travelers were Professor Chaim Weizmann, President of the Zionist World Organization, discoverer of trinitrotoluol, and head of the British Admiralty laboratories during the war; Michael Ussichkin, a member of the Zionist delegation to the Paris Peace Conference and now Resident Chairman of the Zionist Commission in Palestine, and Dr. Benzion Mossinson, President of the Hebrew Teachers' Organization in Palestine.

The party was welcomed at the Battery by thousands of fellow-Jews who had waited there for hours.

The crowds were packed deeply along the Battery wall, waving Jewish flags of white with blue bars, wearing buttons with Zionist inscriptions, and cheering themselves hoarse as the police boat John F. Hylan drew near.

Dozens of automobiles were parked near the landing, and when the welcoming committee and the visitors had entered them they started uptown to the Hotel Commodore, preceded by a police escort. They turned into Second Avenue, where the sidewalks were lined nearly all the way uptown with thousands who waved hands and handkerchiefs and shouted welcome to the visitors.

Professor Einstein was reluctant to talk about relativity, but when he did speak he said most of the opposition to his theories was the result of strong anti-Semitic feeling. He was amused at attempts by reporters to get some idea of his theory by questioning him, and he did his best to make his answers as simple as possible. He spoke through an interpreter.

A Theory of Space and Time

The interview took place in the Captain's cabin, where Professor Einstein was almost surrounded by seekers after knowledge. He was asked to define his theory.

"It is a theory of space and time, so far as physics are concerned," he said.

"How long did it take you to conceive your theory?" he was asked.

"I have not finished yet," he said with a laugh. "But I have worked on it for about sixteen years. The theory consists of two grades or steps. On one I have been working for about six years and on the other about eight or nine years.

"I first became interested in it through the question of the distribution and expansion of light in space; that is for the first grade or step. The fact that an iron ball and a wooden ball fall to the ground at the same speed was perhaps the reason which prompted me to take the second step."

He was asked about those who oppose his theory, and said:

"No man of culture or knowledge has any animosity toward my theories. Even the physicists opposed to the theory are animated by political motives."

When asked what he meant, he said he referred to anti-Semitic feeling. He would not elaborate on this subject, but said the attacks in Berlin were entirely anti-Semitic.

Dr. Einstein said the theory was a step in the further development of the Newtonian theory. He hoped to lecture at Princeton on relativity before he left the country, he said, as he felt grateful to the Faculty of Princeton, which was the first college to become interested in his work.

Poses for Moving Picture Men

As the questioners gave up their attempts to seek further elucidation of the Einstein principles, the professor laughed and said:

"Well, I hope I have passed my examination."

Professor Einstein's interview came soon after he had escaped the moving picture men. As they ground away at their machines, ordering him about, he seemed at first bewildered, then amused. He posed with other members of his party and with Mrs. Einstein for nearly half an hour, and then almost ran away, shaking his head in exasperation and refusing to do any more.

"Like a prima donna," he exclaimed.

"He does not like to be what you call it, a showcase," said Mrs. Einstein. "He does not like society, for he feels that he is on exhibition. He would rather work and play his violin and walk in the woods."

"Do you understand his theory?" Mrs. Einstein was asked.

"Oh, no," she said, laughing, "although he has explained it to me so many times. I understand it in a general way, but in its details it is too much for a woman to grasp. But it is not necessary for my happiness."

Dr. Einstein was an inspirational worker, she said. When he was engaged in some problem, "there was no day and no night," but in his periods of relaxation he went for weeks without doing anything in particular but dream and play on his violin. Whenever he became weary in the midst of his work he went to the piano or picked up his violin and rested his mind with music.

"He improvises," she explained. "He is really an excellent musician."

[Then come passages on Weizmann's impression on Einstein and on Einstein's preferences in music, which have already appeared on p. 10.]

Professor Weizmann also is accompanied by his wife. He and the other Zionist visitors, during their visit of several weeks, will endeavor to interest American Jews in the Zionist movement and obtain money and moral support for both the national Zionist idea and for the university.

Dr. Weizmann Explains Mission

"It is a great satisfaction to me as President of the Zionist Organization to find myself for the first time in the United States," said Dr. Weizmann. "The cause of the Jewish national home in Palestine has from the first appealed to the generous instincts of the American people and owes much to the sympathetic support it has consistently received from leaders of public opinion in the United States.

"Our primary object is to confer with the American Zionists who have, under the distinguished leadership of Justice Brandeis, Judge Mack and other representative American Jews, rendered invaluable services to the Zionist movement during the past few critical years. In the task of reconstruction in Palestine for which the time has now arrived, it is confidently expected that the American Zionists will play an equally conspicuous and honorable part. In this connection we hope to enlist the active interest of American Jews in the Keren Hayesod, or Foundation Fund, the central fund for the building up of the Jewish National Home, to which Jews throughout the world are being called upon to contribute to the utmost limit of their resources.

"Professor Einstein has done us the honor of accompanying us to America in the interest of the Hebrew University of Jerusalem. Zionists have long cherished the hope of creating in Jerusalem a centre of learning in which the Hebrew genius shall find full self-expression and which shall play its part as interpreter between the Eastern and Western worlds. Professor Einstein attaches the utmost importance to the early inauguration of the Jerusalem university and is prepared when the time arrives personally to associate himself within its activities—a course in which there is reason to hope he will be followed by other Jewish scholars and scientists of world-wide reputation."

Einstein to Work for University

Professor Einstein will devote most of his time while here to advocating support of the university by American Jews.

[The *New York American,* April 3, p. 5, quoting the same statement, introduces it with the following two paragraphs:]

Through Captain Tulin, who acted as interpreter, Professor Einstein said:

"The object of my visit to America is to assist the Zionist organization to secure the support, both material and moral of American Jewry for the Hebrew university of Jerusalem.

"The establishment of such a university has been for a long time one of the most cherished plans of the Zionist organization," he said. "But for the outbreak of the war it would have materialized in 1914, when a site was actually purchased on the Mount of Olives. In 1918 the foundation stone was laid by Dr. Weizmann. Since then the university site has been extended and a building purchased in which it will be possible for a beginning to be made. There is also a library of 30,000 volumes which is rapidly growing.

"Plans have been worked out both for the complete university of the future and for a comparatively modest beginning. The time has now come to insure the immediate realization of the latter. Such is the importance attached by the Zionist Organization to the spiritual values in the Zionist national home that even at this moment, when the organization is faced with tremendous tasks of immigration and colonization, and is concentrating all efforts upon the Palestine Foundation Fund, an exception is made in favor of the university to which a special branch of the fund is devoted.

"I know of no public event which has given me such delight as the proposal to establish a Hebrew university in Jerusalem. The traditional respect for knowledge which Jews have maintained intact throughout many centuries of severe hardship made it particularly painful for us to see so many talented sons of the Jewish people cut off from higher education and study, and knocking vainly at the doors of universities of Eastern and Central Europe.

Home for Spiritual Life

"Others who have gained access to the regions of free research only did so by undergoing a painful, even dishonoring, process of assimilation which crippled [the free and natural development of the spiritual character of our people, the *New York American* adds] and robbed them again and again of their cultural leaders. The time has now come when our spiritual life will find a home of its own. Distinguished Jewish scholars in all branches of learning are waiting to go to Jerusalem, where they will lay the foundation of a flourishing spiritual life and will promote the intellectual and economic development of Palestine.
[The *New York American* adds the following sentence:]

"Our Hebrew University of Jerusalem will become the spiritual centre not only of Palestine but also of the Jewish people scattered all over the world.

"Notwithstanding the crude political realism of our times and the materialistic atmosphere in which it has enveloped us, there are visible none the less glimmerings of a nobler conception of human aspirations, such as were expressed in the part played by the American people in world politics. And so we come from sick and suffering Europe with feelings of hope, being convinced that our spiritual aims will command the full sympathy of the American nation and will receive enthusiastic approval and powerful support from our Jewish brethren in the United States."

The Zionists were met down the bay by a delegation from the Mayor's committee of welcome, Captain Abraham Tulin, who served as American liaison officer with General Mangin's army in the war; Dr. Schmarya Levin, who was member of the first Russian Duma and of the Cadet Party in Russia; and Magistrate Bernard Rosenblatt. They were delayed by the quarantine examination and were not able to board the Rotterdam until nearly 1 o'clock. On the way up the bay they had lunch with Professor Einstein, Professor Weizmann and others in the party, and remained with them on the ship until sundown. As it was the Sabbath their religion prevented them from leaving until that time.

Crowd Waits Four Hours at Pier

At the pier were several hundred welcomers, although the ship was more than four hours late in reaching her pier. They gave the Zionists a rousing welcome before they went aboard the police boat John F. Hylan, which landed them at the Battery. The boat flew the Jewish flag in honor of the party. On board were L. Lipsky, Secretary of the Zionist Organization of America; L. Robison of the National Executive Committee; B. G. Richards, Secretary of the American Jewish Congress; M. Rothenberg, Chairman of the American Jewish Congress; J. Fishman, managing editor of *The Jewish Morning Journal;* W. Edlin, editor of *The Day;* Rabbi M. Berlin; David Pinski, editor of *Die Zeit;* John F. Sinnott, Secretary to Mayor Hylan; Henry H. Klein, Commissioner of Accounts; Judge Gustave Hartman, the Rev. H. Masliansky, Judge Jacob S. Strahl and many others.

An official meeting of welcome will be held at the City Hall on Tuesday, at which Mayor Hylan, Frank L. Polk, George W. Wickersham, Magistrate Rosenblatt, Professor Einstein and Professor Weizmann will speak.

Among those on the Committee of Welcome are Nathan Straus, Arthur Brisbane, Chancellor E. E. Brown, Judge Benjamin Cardoza, Abram I. Elkus, James A. Foley, F. H. LaGuardia, Justice Samuel Greenbaum, William D. Guthrie, Mrs. William R. Hearst, Adolph Lewisohn, Alfred H. Smith, Leon Kaimaky, Judge Otto A. Rosalsky, Benjamin Schlessinger, Oscar S.

Straus, Senator Nathan Straus Jr., Marcus Loew, Dr. Bernard Flexner, Colonel Robert Grier Monroe, Herman Bernstein, Samuel Koenig and George Gordon Battle.

A meeting also will be held at the Metropolitan Opera House on April 10. Professor Einstein will not touch on relativity at these meetings, but it is expected that before he leaves the city he will speak before some scientific gathering at which he will discuss his discovery.

The New York Times, April 3, pp. 1, 13.

Other members of the delegation were Solomon Ginzberg (later Shlomo Ginossar), son of the famous Jewish philosopher Asher Ginzberg (later Achad Ha'am) and secretary of the University Fund and (on this occasion) Einstein's personal secretary,; Gerson Agronsky, head of the Zionist Commission Press Bureau; and Weizmann's personal assistant, Leonard Stern. Shmarya Levin joined the mission in New York.

"Our immediate impression of the reception awaiting us in America was far from heartening," Vera Weizmann remembers. "Very few of the 'notables' were there to greet us; but thousands of Jews from Brooklyn and the Bronx came on foot to the harbour to welcome us, and we stayed there for a few hours until Shabbath was over."[6]

The makeup of the delegation shows that the "notables" missing were of the so-called Brandeis group, the top leaders of the Zionist Organization of America (ZOA). Nevertheless, there was Julian Mack, president of the ZOA, and Louis Lipsky, Weizmann's strongest ally in his campaign for the Palestine Foundation Fund (*Keren Hayesod*) in America.

This was the first time that Weizmann was mysteriously referred to as the "inventor of TNT." "'It was in vain that I systematically and repeatedly denied any connection with, or interest in, TNT'—Weizmann writes.—'The initials seemed to exercise a peculiar fascination over journalists, and I suppose high explosive is always news.'"[7] (The solution of the "mystery" is on p. 28.)

My sense is that the reporters noted a significant asymmetry between Einstein's and Weizmann's scientific achievements and wanted to attribute a discovery to Weizmann as important as Einstein's relativity.

According to the *New York Call* (see below), Einstein's and Weizmann's interviews were given on the motorboat *John F. Hylan* on the way from the steamer to the shore and not on board the *Rotterdam*.

The invitation by Princeton University to deliver a series of lectures there in the winter of 1920 was sent to Einstein on October 1, 1920, through Professor Luther P. Eisenhart, who as a specialist in the field of differential geometry was able to appreciate Einstein's theory of general relativity in its full depth.[8]

Einstein Here with Weizmann to Aid Zionism

Only Opposition to Theory of Relativity, the Professor Says, Came from Politicians

One of the world's great scientists, Professor Albert Einstein, author of the theory of relativity, arrived on the Rotterdam yesterday as a member of the delegation which has come to this country in the interests of Zionism. With him were Professor Chaim Weizmann, president of the World Zionist Organization; M. M. Ussischkin, a member of the international Zionist committee, and Ben Zion Mossesohn, a leading educator of Palestine.

The delegation, with Mrs. Einstein and Mrs. Weizmann, were met at the pier in Hoboken yesterday after sundown by a committee of prominent Jews aboard the police tug John F. Hylan. They were landed at the Battery. Here, amidst a flurry of cheers and flags, a small parade formed and a chain of taxicabs whisked the visitors to the lower East Side which turned out en masse to greet them, to the Hotel Commodore.

Crowded in the little cabin of the John F. Hylan, Professor Einstein talked to reporters in the trip from Hoboken to the Battery and described the long years of effort through which he had worked out his theory. He spoke only German, and his words were interpreted by everyone in the vicinity who could understand him.

Talks in German

The distinguished physicist is a broadly built man who looks more the artist than the scientist. He has a fine, graying head with the hair falling loosely back from his high forehead when he removes his hat. His coloring is warmly brown, the features prominent, the eyes and mouth friendly. He seemed to take a childlike amusement in the whole process of questioning and scribbling and interpreting that was going on before him, but all the while felt about him the large simplicity of the scientist who deals with vast spaces and elemental things.

When first asked to explain his theory of relativity, Professor Einstein, who has previously said that there were only a few people in the world capable of understanding it, laughed gayly. Then he began to put it into as exact German as his hearers could understand.

"It is a theory of space and time as far as physics is concerned, which leads to a definite theory of gravitation," he said. "The theory consists of two steps."

Professor Einstein explained that he had worked on the first step for seven years and on the other for eight or nine years. He was drawn to this research through the expansion of light in space and through the fact that iron and wooden balls fall to the ground with the same degree of acceleration in spite of the law of gravitation.

"I believe that this theory is a step further than Newton's law of gravitation, in regard to definiteness as well as to method," he continued.

Opposition Political

Asked about the opposition which had manifested itself after his announcement of the theory of relativity, Professor Einstein said it had been purely "political." No real scientist has any prejudice against his theory. What opposition there was the noted scientist explained partly on the ground of anti-Semitism, and partly on the ground of "politics."

Professor Einstein said he had come to this country to secure help for the establishing of a Zionist university in Jerusalem. He hopes to lecture in it some day when the dream is realized. At present his plans are uncertain, though he expects to lecture at Princeton because it is the first American university that has asked him.

Weizman, Jewish Hero

A little earlier Dr. Weizmann, in the midst of an eager crowd on the rear deck of the boat, explained the purposes of the delegation in visiting America and what the Zionist leaders hope to accomplish in Palestine. Dr. Weizmann, formerly professor of chemistry in the University of Manchester, is both scientist and statesman and is a hero among the Jewish people for his success in the game of diplomatic chess at the peace conference and afterward.

While he was not prepared to say that the British Government favored an independent Jewish state in Palestine, Weizmann said that the ideal of a Jewish national home ultimately meant a state.

1,000 Come to Zion Monthly

Immigrants are coming into Palestine now at the rate of about 1,000 a month, Dr. Weizmann said. Most of them are from eastern Europe—Russia, Rumania, the Ukraine—young, unmarried men who have been through the war. At present there are 100,000 Jews and 500,000 Arabs in Palestine, he said. Three languages are used officially—English, Hebrew and Arabic.

The committee appointed by the Mayor to greet the visitors was headed by City Magistrate Bernard A. Rosenblatt. Included in the representatives from Jewish organizations who were on board the John F. Hylan were Commissioner of Accounts Henry H. Klein; Judge Julian W. Mack, president of the Zionist Organization of America; David Pinski, editor of *Die Zeit*; Judge Gustave Hartman and others.

The City Delegation

The Zionist delegation boarded the tug from the Lackawanna pier in Hoboken. A committee composed of Judge Hartman, Magistrate Rosenblatt, Alderman Louis Zeltner, Louis Lipsky, B. Zuckerman and Captain Abraham Tulin, with roses for the ladies of the party, went ashore to greet the visitors and escorted them back to the boat.

Returning to the Battery, the delegation was greeted with cheers and flag waving by a crowd of several hundred persons who had gathered in Battery Park. While searchlights from harbor craft played over the water the visitors stepped ashore and walked through an archway fairly papered with small flags, to the waiting automobiles.

Before setting out Magistrate Rosenblatt spoke a few words of welcome to the visitors in behalf of Mayor Hylan. Then the procession started, while a band struck up the Hatikvah or Jewish national anthem, and piled up beyond the darkness the lights of the city glittered with infinite allurement.

Escorted by a group of Jewish legionaires who fought under General Allenby in Palestine, the automobiles swept through the lower East Side, by

Zionist Leaders Greeted by Vast Throng at Their Arrival in New York.

"Vast throngs of New York Jews turned out to greet Professor Chaim Weizmann, discoverer of T.N.T. and president of the World Zionist Organization, and Professor Albert Einstein, famous savant, upon their arrival here on the S. S. Rotterdam. This photograph, taken aboard the liner, shows Menaches M. Ussishkin, chairman of the Zionists of Palestine; Dr. Chaim Weizmann, Mrs. Weizmann, Professor and Mrs. Albert Einstein, and Ben Zion Mossinson, principal of the Jaffe College."

The Jewish Independent, April 8, p. 1.

Central Zionist Archives, Jerusalem, courtesy AIP Emilio Segrè Visual Archives.

way of East Broadway, Clinton, Delancey, Forsyth streets and Second avenue, to the Commodore. Crowds displaying Zionist flags lined the streets and cheered as the cavalcade passed.

The New York Call, April 3, pp. 1, 6.

Here one finds the first of many instances of journalists wrestling with relativity: mind those balls falling with the same acceleration "in spite of the law of gravitation"!

After having read a description of Einstein's appearance, the reader might like to read a characterization of Weizmann, too: "His sheer physical presence was arresting, for one thing. The brow of his massive bald head was finely etched with veins, his eyes were piercing, his moustache and goatee elegant, his clothing always superbly tailored."[9] A description of their wives is on pp. 124–128.

It seems that Weizmann succeeded in convincing Einstein to accept a future appointment at Hebrew University, although Einstein's "promise" sounds rather cryptic; a few months earlier he had definitely rejected this possibility, if only for the reason that he could not speak Hebrew. In fact, he never lectured there.

The *New York American* gives other details of the events below.

City Welcomes Prof. Einstein and Weizmann

Discoverer of Theory of Relativity and President of Zionists of World Hailed by Motor Fleet—Big Reception Tuesday—Hylan Committee Heads Welcomers Down Bay—Distinguished Jews Here to Aid Palestine

A float of motorboats, loaded with men and women, welcomed Dr. Chaim Weismann, president of the Zionist Organization of the World, and Professor Albert Einstein, discoverer of the theory of relativity, when they arrived here from Europe yesterday.

The welcomers were headed by Mayor Hylan's Committee of Welcome appointed to greet the world-famous Jews.

Hundreds of automobiles were waiting in Battery Park when the Police Boat John F. Hylan landed the distinguished visitors at 7 P.M.

These fell into line behind the Welcome Committee machines and escorted the visitors to the Hotel Commodore, the parade moving uptown through the East Side.

Thousands in Bowery Cheer

A noisy and picturesque feature of the welcome at Battery Park was a contingent of the Jewish Legion, which fought under General Allenby in the Holy Land campaign, who had a fife and drum corps with them. Thousands cheered the oars as they passed along the Bowery and up Second avenue.

Einstein in motorcade on the occasion of his arrival in New York City.
Courtesy Brown Brothers.

The two Zionists and their party are visiting this country to enlist support for the Jewish national home and college for Jews in Palestine.

Early yesterday the Mayor's Committee of Welcome, headed by Magistrate Bernard A. Rosenblatt, Captain Abraham Tulin and Dr. Shayrm Levine, went down the bay to meet the steamship Rotterdam. They waited until 1 o'clock before the big liner was permitted out of Quarantine.

As soon as the committee piled aboard the ship from the coastguard cutter Guthrie, Mrs. Weizmann effusively greeted Dr. Levine, who was one of the original members of the Russian Duma and expressed her happiness at being in the United States.

Refuse to Break Sabbath

Though the ship docked during the afternoon, Dr. Weizmann and Professor Einstein refused to break the Jewish Sabbath and travel on Saturday. They remained aboard the vessel at Hoboken until sundown. Then, with their wives and members of their party, they boarded the police boat John F. Hylan, which had been placed at their disposal by Mayor Hylan. They were taken to the Dock Department pier at the Battery, and then, under police escort, to the Hotel Commodore.

[Here follows Einstein's statement on his mission in the United States, which is on pp. 18–19.]

Professor Einstein is a member of the Academy of Science of Berlin, and also a professor at the University of Leyden, Holland. He was born in 1879. In 1902 he entered the Swiss patent office. In 1905 he received a degree of Ph.D. of Zurich. Since 1913, Professor Einstein, though a Swiss citizen, held a privileged position at the Berlin Academy of Science.

His theory of relativity introduced a new scientific conception of space and time and their relation to the physical world, with important consequences for the theory of light and gravitation.

Dr. Weizmann during the war presented Great Britain with the discovery of a method whereby powder could be kept absolutely dry aboard ship. He was born near Pinsk, Russia, in 1874. He was lecturer of chemistry in the University of Geneva, and later reader in bio-chemistry at the University of Manchester.

Accompanying the two famous men are Dr. Michael Ussichkin, chairman of the Zionist Commission, and Dr. Benzion Mossinson, president of the Hebrew Teachers' Association.

New York American, April 3, p. 5.

Here it is not TNT but gunpowder that is associated with Weizmann. This guess is closer to the truth, so it is time to reveal what was behind the TNT "mystery."

Before World War I, Weizmann discovered that a bacterium—later named after him—*Clostridium acetobutylicum* (Weizmann) can turn sugar into acetone and butyl alcohol. When the war broke out, the British Navy was in need of huge amounts of acetone, necessary for the production of smokeless powder, which would enable the warships not to be located by the smoke of their gunfire.

In Manchester the director of research for Nobel Explosives in Scotland met with Weizmann, who told him about his process.

Let me continue with the words of John C. Polanyi, winner of the

1986 Nobel Prize for chemistry. His story is close to what really happened and makes for much better reading than a dry report:

"In no time Weizmann found himself on a train to London to be interviewed by the First Lord of the Admiralty. He explained to His Lordship that the bacterium . . . would readily make the necessary tons of acetone provided that a sufficiently large fermentation and distillation plant could be found. The First Lord, a Mr. Winston Churchill, explained kindly to Weizmann that he knew some distillers, and though he would not wish to interfere with the supply of whisky he could get Weizmann a gin factory. Weizmann thereupon became chief scientist of the Nicholson Gin Company, and an ally of a politician with an interest in a Jewish National Home in Palestine."[10]

After the war, the other end product of the fermentation process, butyl alcohol, also became of high value, being an important component of auto lacquers.

With this process Weizmann initiated microbiological methods in industry—an achievement of significant technological importance.

Einstein Explains Relativity Theory

German Professor Says Discovery Carries Newton's Law a Step Forward—Work Not Yet Finished—Wife of Scientist Declares the Whole Matter Is Rather Dim to Her

NEW YORK, April 2.—Before leaving Berlin Prof. Albert Einstein voiced a prayer that he should not be interviewed in America. But this afternoon he answered all the questions that a groping lot of newspapermen and newspaper women could think of. And one could tell from his chuckling and his willingness to stay until the "worst was over" that he enjoyed it.

Thus the originator of the theory of relativity, which several persons profess to understand and a few undoubtedly do, although Mrs. Einstein, who is with her husband, says it's rather dim to her, surrendered, as they all do, to a great American tradition. Incidentally, he made captive his interviewers at the same time.

Prof. Einstein comes here as one of the Zionist commission, whose purpose is to interest American Jews in the restoration of Palestine, and to tell them how it may be done. His special interest is in the establishment of a university in Jerusalem. The head of the delegation is Dr. Chaim Weizman, formerly professor of chemistry in the University of Manchester, now president of the World Zionist Organization. The two others are M. M. Ussischkin and Ben Zion Mossesohn, head of the Jewish High School at Jaffe.

These four, whose arrival has been feverishly awaited by a multitude of Americans, arrived on the Rotterdam. At the Battery the Zionists were met by thousands of men and women and a company of Jewish legionaries who fought under Allenby in Palestine.

"Will you define in simple terms your theory of relativity?" was the first question asked.

"It is a theory," the doctor readily replied, "of space and time so far as physics is concerned. It leads to a definite theory of gravitation. That is all one can say in a few words."

"How long were you in arriving at the theory?"

"I am not through yet." He waved his short, curved-stemmed pipe with a gesture indicating much work ahead. Then he added: "I worked for about sixteen years. The theory has two phases. The first one occupied seven years, the second eight or nine."

"And what led you into this field of speculation?"

"I came to it through a study of the phenomena of the expansion of light through space. That was the first phase. The second phase was the observation of the effect of the movement of various bodies. For example, the manner in which they fall to the ground irrespective of their physical nature—the fact that an iron ball and a wooden ball of the same size fall from a given point with the same velocity."

"No man of culture and knowledge has any prejudice against my theory," he said. "Physicists who do oppose it are animated by political reasons in my opinion. That is due primarily to anti-Semitism. I do not care to say more about that."

"Do you believe you have carried Newton's law of gravitation a step further?"

"Yes," he answered, after looking at the questioner rather quizzically, it seemed. "I believe it is a step forward in definiteness as well as in method."

He was asked, of course, about the significance of his discovery that rays of light were slowed up and bent upon entering a field of gravitation. He said: "That merely proves the truth of my theory."

The Public Ledger (Philadelphia), April 3, p. 2.

The *Ledger* puts the time at which Einstein was stormed by the newspapermen in afternoon, that is, when the delegation was on board the *Rotterdam*. I am pretty sure he was interviewed both in the captain's quarters on the *Rotterdam* and on the tugboat.

The man requested to take charge of Einstein's press relations was Louis ("Pop") Popkin. His wife, Zelda Popkin, tells the story of Einstein's encounter with the journalists as follows: "Pop met Albert Einstein and explained to him that he was about to receive a general press interview. He explained what a press interview was. Professor Einstein winced. 'I can't do that,' he protested. 'It's like undressing in public.' Pop reasoned with him and, as always, was convincing. Einstein was introduced to the ship newsmen. 'Professor,' the *City News* representative rushed in where angels had hesitated, 'will you be good enough to tell us in one sentence exactly what your theory of relativity is?'"

"The professor stared, not believing his ears. 'All my life,' he said in German, 'I have been trying to get it into one book. He wants me to state it in one sentence.'

"'Professor, what do you think of America?' the *Tribune* asked.

"'Excuse me, I have not yet arrived in America,' the professor said.

"'Professor, what do you think of American women?' a tabloid reporter asked.

"'Excuse me, I have not yet met with American women.'

"'Professor, exactly how will your theory benefit the man in the street?' the *New York Times* asked.

"The professor looked around for an exit and, like a scared rabbit, he fled. It was a cruel, crude performance. Pop, himself, seemed unhappy whenever he spoke of it."[11]

Even though it sounds more like a folk tale than a word-for-word account, it nicely recreates the scene of Einstein besieged by the journalists.

Einstein Explains 2 Phases of Theory

Famed Scientist, Through Interpreter, Tells of Discovery That Light Rays Are Bent—Goes Newton One Better—"Iron Ball Can't Fall Faster Than Wooden One," He Asserts—Pays Tribute to America

Professor Albert Einstein, eminent scientist, who has caused worldwide discussion by theories of relativity and bent light, explained his discoveries last night after a cordial welcome to America.

Although a Swiss citizen, Professor Einstein spoke in his native German, and it was difficult for him to convey his technical discussion to lay hearers through an interpreter. Equally difficult was it for him to explain the antagonism his theory directed toward him in Germany, forcing him to leave Berlin.

His explanation of his science, as translated, follows.

"I worked sixteen years on my theory, and I am sure of my ground. Much attention has been given to discussion of my proposition that light in diffusion is bent. Always we had followed the theory that light did not bend. This part of my theory is not of importance, except as an evidence of the entire theory.

Two Phases of Theory

"There are two steps in my theory. On one I worked for seven years and on the other eight or nine years. I maintain that falling bodies or objects are independent of physical causes.

"An iron ball and a wooden ball fall with the same acceleration. My theory is a step in advance of the Newton theory on the law of gravitation. I believe it an advanced step toward definiteness as well as to improvement of scientific method.

"I hope to tell the American people about my theory from the lecture platform. It is very likely my first lecture will be in Princeton University, it being the first institution to invite me.

"You ask me the reason for the attacks upon me in Berlin. Well, no man of science opposes my theory. Such opposition as there has been has had an entirely political origin.

Scientist Blames Politics

"Even the physicists opposed to my theory are usually animated, in my opinion, by political consideration. Their attitude is largely due to anti-Semitism. The attacks in Berlin were entirely anti-Semitic. Then, too, I was a pacifist."

When it was announced through the interpreters that the interview was at an end Professor Einstein looked happy and remarked:

"I hope I stood my examination well."

Thereupon the man with a great theory to discuss with the scientists of the world went to the bow of the police boat Hylan and exclaimed at the wonders of the city. He exclaimed:

"It is magnificent! Grand America! Grand New York! The one country that shelters my people!"

He chatted laughingly with those who could understand his tongue, and they said he was talking boyishly about the treat in store for him in this country.

New York American, April 3, p. 3.

Please note that Einstein spoke German, "although a Swiss citizen." This led to confusion for some journalists who wondered not why Einstein avoided French, Italian, or Romansh, one of the non-German languages the Swiss use, but why he did not use "the" Swiss language. Einstein had two reasons for using German: first, it was his mother tongue, and second, there is no "Swiss" language in the sense that, for example, French people speak the French language.

The journalists' grasp of physics, however, is getting better. Here the balls no longer fall "in spite of the law of gravitation"; now they are only "independent of physical causes."

Another confusing expression is "light in diffusion." Let's try not to blame the journalists. Einstein must have used the German expression *Lichtausbreitung* (light propagation), and, because the literal translation of the word *Ausbreitung* is "spread out," it was translated by the casual interpreter as "diffusion," which roughly means something going outwards in every direction. Maybe the falling balls also lost their physical causes somewhere between Einstein and his interpreter.

The enthusiastic exclamation at the end of the article sounds very non-Einsteinian. Maybe it was difficult to understand him because he was being tongue-in-cheek.

Einstein's Music Excels Science in Wife's Opinion

Professor Albert Einstein has commenced one of the greatest pro and con discussions in the world of science, with his theories of relativity of time and space, but to his wife he is just a plain, music-loving husband.

After the arrival of the Rotterdam, which brought Professor Einstein to port yesterday, Mrs. Einstein was asked to describe her husband's famous theory. Laughingly she raised her hands and exclaimed:

"I have known my husband for years, but to tell the truth, I don't understand about the subject of which all the world is talking. Scientists may—but I—never! I would rather hear him play his violin or his piano than hear him discuss scientific problems.

"I am afraid that he works too hard. But he calls it mere play. Often I do not hear a word from him for days and days. He knows not the difference between night and day. When a problem is put up to him in the work he loves he works days and nights without sleep or food. Then to rest his tired body he will turn to the violin or the piano.

"Then I sit and listen to him by the hour—but science—never!"

New York American, April 3, p. 3.

The next article was composed of the two previous ones, but with further details added.

Enthusiastic Reception Is Accorded Zionist Delegation

Dr. Weizmann, Professor Einstein and Other Members of Party Greeted by Thousands—New Yorkers Welcome Visitors—Surging Crowd Meets Delegation on Arrival Saturday—Dr. Einstein Issues Statement Explaining Purpose of Visit—Desires Support for Hebrew University of Jerusalem

NEW YORK, April 5.—(J. C. B. Service)—Few individuals have ever caused such a stir in the communal life of Greater New York as did the arrival of Dr. and Mrs. Chaim Weizmann, Dr. and Mrs. Albert Einstein, and Menachen Mendel Ussishkin. Thousands gathered at the pier early on Saturday to await the arrival of the Zionist delegation, which did not actually come ashore until after sunset. On landing on New York soil these noted Jews and their party were literally swept off their feet by the surging crowd.

On board the S. S. Rotterdam, on which the delegation arrived, the leaders had previously met and entertained a committee representing the Mayor and citizens of New York and other committees representing the Jewish community of the Greater city. The delegation also acknowledged the call of numerous newspaper men, as well as the host of cameramen that are always on hand to greet those of any distinction. After the Star Spangled Banner and Hatikvah were sung, it was some time before the cheering subsided and the police succeeded in clearing the road for the procession. In the course of the evening and throughout Sunday members of the delegation which includes also Dr. Benzion Mossinsohn, president of the Hebrew Teachers of Palestine, received numerous visitors and delegations representing various institutions in as well as out of town. In the afternoon the delegates were at lunch with Mr. and Mrs. Samuel Untermeyer at his home.

Professor Einstein who made use of an interpreter made it clear at the outset that he is a Swiss and not a German. He was merely born in Germany and is at present professor and a member of the Academy of Science in Berlin. He is forty-two years old, has a decided bent for music and is greatly addicted to the use of a pipe.

[Then comes the statement reproduced on pp. 18–19.]

The Jewish Independent, April 8, pp. 1, 8.

> Whenever the journalists failed to get a short, easy-to-understand account of relativity from Einstein, they turned to Weizmann with the same request, saying, "But you're a scientist, too." In the end, "in sheer desperation, we took refuge in an inconspicuous cabin and waited till it was time to go ashore," remembers Weizmann. After a long and noisy procession through the Jewish section of New York, the delegation arrived at the Hotel Commodore shortly before midnight.[12]
>
> To land at the Battery and not on the wharves above Greenwich Village was characteristic of grand receptions: "This prerogative of fame began perhaps with Lafayette, who was officially received at the Battery in 1824. Of late there have been any number of heroes, and one or two heroines, welcomed at the Battery, celebrities to whom the Mayor of New York, in a snowstorm of ticker-tape, gave the freedom of the city."[13] In this case the snowstorm of ticker-tape was replaced by little American and Zionist flags. Otherwise everything went as usual, even the freedom of the city—although with a hitch.
>
> The last program of the day was a small reception.

An impromptu reception was held in the evening in one of the grill rooms of the hotel. Dr. Weizmann, accompanied by the members of the Delegation, appeared and was most enthusiastically received.

Mr. Louis Lipsky, Secretary of Organization, for the Zionist Organization of America, welcomed the leaders on behalf of the Zionists of America and Dr. Weizmann responded briefly, laying especial emphasis upon the duty devolving upon American Jewry to take full advantage of the political facilities which were secured for the restoration of the Jewish Palestine. Speaking for Professor Einstein and Mr. Ussischkin, Dr. Weizmann, the head of the Delegation, expressed the hope that there would be many opportunities for meetings with the Zionists of the country and that before their departure, after their necessarily limited stay in the country, definite progress toward the achievement of their object would be made.

Yidishes Togeblat—The Jewish Daily News, April 5.

Weizmann and Einstein Discuss Plans

Special Interviews by Jewish Correspondence Bureau

Questioned as to the purpose of his visit, Professor Weizmann declared that he is bringing with him important plans for the rebuilding of Palestine, plans which can only be realized with the entire help of the Jews of America. The European Zionists, Dr. Weizmann said, are at present working under extreme difficulties and not only is it essential that American Zionists should concentrate upon the work at hand with all their energies but it is his hope that they will also accord the European workers all necessary encouragement.

Pending the meeting of the next congress, Dr. Weizmann said, the newly-composed executive has reached an agreement regarding all activities. The congress will be called at an early date and preparations for its meeting have already been started. The organization is now looking around for a suitable town and all formalities with regard to the actual elections have already been defined. Asked as to whether he intended to ask American Zionists to enter the executive Dr. Weizmann replied that at this moment this would not be practicable since none of them appear to be in a position to proceed to Europe before June, about which time the next meeting of the Greater Actions Committee will be held.

Dr. Weizmann stressed the importance of the work to be done by the Economic Council. The Economic Council, he said, will shortly be registered as a company and although it will work in close touch with the Zionist organization, it will remain entirely independent of the organization. The organization will, however, have two representatives on the Council both of whom will be appointed by Dr. Weizmann himself. Similarly, the ICA will have two of its representatives on the Council.

The Council will not collect any funds from the public at large but will concentrate on financing definite projects. It is probable that the Council will also reorganize the existing Zionist banks. Of its income the Keren Hayessod will hand over to the Council that third which is intended for investments. Questioned as to whether he intends to remain at the helm of the Zionist movement Dr. Weizmann said that that was a matter dependent entirely upon the future, adding that he retained the leadership at present upon the sole condition that the next Zionist congress be called forthwith.

Questioned regarding the present expenditure in which the Zionist Organization was involved in Palestine, Professor Weizmann stated that the budget was kept within the limits established by Messrs. Julius Simon and Nehemia De Liema, the representatives on the executive who expressed the sentiments of the American Zionist leaders. This budget provides for an expenditure of £14,500 but does not cover immigration. For March £25,000 has been forwarded to Palestine, 10,000 of which will be allotted to immigration work.

Professor Albert Einstein, the brilliant scientist who accompanies the Zionist Delegation chiefly for the purpose of interesting American Jews in the Hebrew University which is being planned for Jerusalem, when approached, dwelt upon the possibilities of that institution. Einstein believes that the influx of Jewish students into Palestine from every part of the world will convert the Jerusalem University into a Jewish spiritual centre.

The Jerusalem University, Einstein said, is not to be compared with any of the developed universities in the western countries. This institution will have to adapt itself to the necessities created by the natural conditions in Palestine. Of prime importance will be an agricultural institute, and possibly also chemical and biological institutes to fight diseases and epidemics. These institutes would be in a position to utilize the existing agricultural experimental stations. An essential branch of the work of the university would be an oriental institute to investigate the country, its historical possessions and the local languages.

Einstein declared that he looks upon the early inauguration of the university as of the highest importance and in order that it should not be exposed to any arbitrary and private initiative the Keren Hayessod must lay its basis. Lectures, Dr. Einstein said, would not be urgent at the outset in view of the fact that the Jewish population is still comparatively small, but the university will grow with the growth of the population and its teaching activities will be increased accordingly.

The establishment of a cultural centre in Palestine, Dr. Einstein believes, will offset Jewish assimilation and will fill Jews everywhere with joy and pride. Dr. Einstein expresses himself as entirely opposed to the idea of accustoming the future students of the Palestine university to one-sided spiritual professions in the manner in which it is at present done in European universities.

Yidishes Togeblat—The Jewish Daily News, April 3.

Weizmann, Ussischkin and Einstein

Brief Sketches of the World Zionist Leaders

Interest in the international Zionist figures who have just arrived to this country is greater than mere curiosity as to the cause of their visit. American Zionists are anxious to know something about the men who have done so much for Zionism. A brief outline of their work is hereby given.

Dr. Weizmann, president of the World Zionist Organization, is a product of the Zionist Movement, having early been active in the Chovevei Zion. A Talmudist and Hebraist from Russia, in early life he went to Switzerland in whose Universities he studied chemistry. He was a lecturer in Chemistry at the University of Geneva and in 1905 he was appointed professor of chemistry at the University of Manchester.

Dr. Weizmann had been doing Zionist work all during the period, and he was known in Zionist ranks as the advocate of a project to found a Jewish University in Jerusalem. After the Balfour Declaration, when England had conquered Palestine, and he had interested a number of people in the University project, he acquired land for the University on Mt. Scopus and laid there the foundations of the Hebrew University, the science branch of which it is expected will be opened by the end of this year.

He was head of the Zionist Commission to Palestine for some time, and to him is ascribed most of the credit for the Zionist diplomatic successes of the past few years, among them the procuring of the San Remo Decision.

M. M. Ussischkin has been an ardent Zionist since his youth when, while yet in school, he became one of the members of the "Bilu," the Jewish National Students' Organization which formed the first Jewish colony in Palestine. Afterwards he was one of the Organizers of the Students' Chovevei Zion and Ben Zion Conference at Drusgenik, Government of Grodno, and in 1890 he was one of the founders of the Odessa Association for Aiding the Jewish Colonization of Palestine.

After the Palestinian scheme had been transformed into the present political Zionist movement, Ussischkin became one of the most ardent followers and collaborators of Herzl with whom he began a correspondence in 1896, and since then he has been one of the most energetic propagandists of Zionists among the Russian Jews. Ussischkin has taken part in all the Zionist Congresses excepting the 6th, and was one of the members of the Zionist Actions Committee. In 1903 he was sent to Palestine by this committee and by the Chovevei Zion to purchase land for new colonies and to organize the colonists and other Jews of Palestine.

Ussischkin has been a member of the Zionist Commission, and until his departure for America was head of the Commission.

[The newspaper here repeats Einstein's curriculum vitae, which is on pp. 11–12.]

Yidishes Togeblat—The Jewish Daily News, April 3.

On November 2, 1917, British Foreign Minister Arthur J. Balfour sent a letter to Baron Edmond de Rothschild in which he informed him that "His Majesty's Government views with favour the establishment in Palestine of a national home for the Jewish people, and will use their best endeavours to facilitate the achievement of this subject."[14]

Sunday, April 3

Einstein Sees End of Time and Space

Destruction of Material Universe Would Be Followed by Nothing Says Creator of Relativity—Theory "Logically Simple"—Science Burdened Hitherto by Complicated Assumptions, He Asserts—Entertains Many Visitors

Those persons who may have comforted themselves with the reflection that no matter if the worst happened, and everything material in the universe were destroyed, there would still be time and space in which lonesome and expatriated spirits might wander, did not take Professor Albert Einstein into consideration at all. He said jocularly yesterday, having supposititiously destroyed matter by a wave of his hand in which was clutched the omnipresent briar pipe, that under his theory even time and space would then cease to exist.

Professor Einstein said it with a smile, attempting to convey by humor to his puzzled interlocutors a conception of relativity which they had failed to grasp in more abstruse definitions. He is becoming use to having persons ask him to explain what relativity means.

It really isn't as bad as it seems, he explained, and exploded the accepted story that he had said only twelve men in the world were capable of understanding it. He thinks most scientists understand his theories, and added that his students in Berlin understand them perfectly. No theory can be said to be susceptible of absolute proof, he added, and mentioned that an American scientist, St. John, is now conducting experiments which seem to give results at variance with the Einstein theory.

"The two theories, that of St. John and my own, have not yet been brought into harmony," said Professor Einstein. "The subject dealt with is that of the wave lengths in the spectrum. It is impossible at the present stage of the experiments to say what the result will be."

He mentioned this experiment, he said, because he did not wish to seem to claim infallibility for the theory of relativity, and desired to be perfectly fair in his attitude toward the theories of others. The views of Professor Charles Lame Poor, Professor of Astronomy at Columbia, who has said that the Einstein theories cannot be proved and that it is possible to explain all physical phenomena, even the irregularities of Mercury, by Newtonian law, were called to Professor Einstein's attention.

"I did not see Professor Poor's statement," he said. "In a certain sense you can say of any scientific theory that it cannot be proved. No theory can be proved absolutely. Every theory tries to explain certain facts, and it is acceptable in so far as those facts fit into the general conception of a theory. No facts can be said to be explainable by only one theory, and in that sense one might say that the theory cannot be proved."

Professor Einstein was rather puzzled to account for the public interest in his conception of time and space, and said the public attitude seemed to call for a psychologist who could determine why persons who are not generally interested in scientific work should be interested in him.

"It seems psycho-pathological," he said with a laugh.

When it was suggested that perhaps people were interested because he seemed to give a new conception of the universe, which next to the idea of God, has been the subject of the most fascinating speculations of the mind, he agreed that such might be the case.

"The theory has a certain bearing in a philosophical sense on the conception of the universe," he said, "but not from the scientific point of view. Its great value lies in the logical simplicity with which it explains apparently conflicting facts in the operation of natural law. It provides a more simple method. Hitherto science has been burdened by many general assumptions of a complicated nature."

Two of the great facts explained by the theory are the relativity of motion and the equivalence of mass of inertia and mass of weight, said Professor Einstein.

"There has been a false opinion widely spread among the general public," he said, "that the theory of relativity is to be taken as differing radically from the previous developments in physics from the time of Galileo and Newton, that it is violently opposed to their deductions. The contrary is true. Without the discoveries of every one of the great men of physics, those who laid down preceding laws, relativity would have been impossible to conceive, and there would have been no basis for it.

"Psychologically it is impossible to come to such a theory at once, without the work which must be done before. The four men who laid the foundations of physics on which I have been able to construct my theory are Galileo, Newton, Maxwell and Lorenz."

Whatever the value of relativity it will not necessarily change the conceptions of the man in the street, Prof. Einstein said.

Prof. Einstein then mentioned again the two things which seem so far to have proved the correction of his theory, that of the explanation it affords of the irregularities of Mercury and the observation of the solar eclipse, which showed that "rays of light in passing a large body like the sun are deviated by its attraction. This I foretold," he said, "and my prophecy was proved by British scientists."

"The practical man does not need to worry about it," he said. "From the philosophical aspect, however, it has importance, as it alters the conceptions of time and space which are necessary to philosophical speculations and conceptions. Just as a joke, however, and not to be taken too literally, it has this effect on any thought of the universe. Up to this time the conceptions of time and space have been such that if everything in the universe were taken away, if there was nothing left, there would still be left to man time and space. But under this theory even time and space would cease to exist, because they are inseparably bound up with the conceptions of matter."

Professor Einstein is so far from being the usual conception, to use one of his own words, of the average man of science that he has made an unusual impression of geniality, kindliness and interest in the little things of life on those who have come in contact with him. His room was continually filled yesterday with visitors. Hardly for a moment did he let his new pipe, which promises to become as famous as his theory, out of his hand.

"I shall now treat the other pipes with contempt," he said, caressing it.

With his wife and Professor and Mrs. Chain Weizmann, President of the Zionist organization, he was the guest at luncheon of Samuel Untermyer. The party motored up Riverside Drive and around the city for a time beforehand, and Professor Einstein was much impressed by his first view of New York.

"My first impression was of the kind and hearty welcome all the members of the delegation received," he said, "and my other of the enormous size of the buildings of New York, which give the city the aspect of a mountain landscape. It is also impressive of health and strength, not only in the people on the streets, but in the buildings, the streets, the means of communication—it all seems so sound and solid."

Delegations of Jews interested in the Zionist movement visited the hotel all day. Rooms were crowded with conferences presided over by Professor Weizmann and other members of the delegation. Four representatives of Canadian Zionists, A. Lewin, H. E. Wilder, Dr. John Shane and C. A. Cowen, called to give their greetings. Later the leaders of the Mizrachi, the orthodox wing of the Zionist organization, called on Professor Weizmann. They included Rabbi Meyer Berlin, Rabbi M. Z. Margulis, B. Biblick and Dr. Bluestone. A delegation from Philadelphia headed by Jacob Ginsburg also called. Messages from all parts of the United States and even from some European countries were received.

The New York Times, April 4, p. 5.

How Professor Einstein Explains Famous Theory

NEW YORK, April 4.—Here is Professor Einstein's own statement on the theory of relativity, as made to a group of New York newspapermen yesterday:

"Until I announced my theory of relativity it was believed generally that if we removed every possible object from the universe time and space would still remain. But now, keeping the theory in mind, I believe that nothing, not even time and space, will remain if we remove all matter from the universe.

"An opinion, and a false one, has been spread widely that the theory of relativity is something contrary to the development of the science of physics from the days of Galileo, Newton, Maxwell and Lorentz. The contrary is true. The theory is a natural outcome of the work of these men.

"European and American astronomers are hard at work on my theory. I have been told that complaint has been made of the difficulty of studying it. It can't be so difficult. My students in Berlin seem to understand it."

The Jewish Independent, April 8, p. 1.

4

City's Welcome

✦

Tuesday, April 5

City to Honor Zionist Leaders Today

Today at the City Hall, the "Freedom of the City" will be conferred upon Dr. Chaim Weizmann, Prof. Albert Einstein and Menachem Mendel Ussischkin, the Zionist leaders who arrived here on Saturday evening.

The Welcome Committee appointed by Mayor Hylan will call upon the distinguished guests at the Hotel Commodore where they are staying and will escort them to the City Hall.

It had at first been arranged that the ceremony should take place in the Aldermanic Chamber, but in view of the wide-spread desire to witness the event, the speeches will be delivered from the steps of the City Hall.

Addresses will be delivered by Mayor Hylan, ex-Secretary of State Frank Polk and Magistrate Bernard A. Rosenblatt.

Dr. Weizmann will respond. This will be the first public utterance of the Zionist leader since his arrival in New York. Next Sunday there will be an official Zionist reception at the Metropolitan Opera House and the following Tuesday evening a popular greeting will be extended to the Zionist leaders at the Sixty-Ninth Regiment Armory.

From the moment of their arrival the Zionist leaders have been besieged by delegations of eager Zionists who have come to greet their leaders and who have thronged the lobbies and halls of the hotel on Pershing Square. After the procession from Battery Park to the Hotel on Saturday, many Zionists lined the streets and for hours waited to catch a glimpse of their leaders from abroad.

[Here follows the paragraph in which an impromptu reception is described; see below.]

Yidishes Togeblat—The Jewish Daily News, April 5.

City's Welcome for Dr. Einstein

With Dr. Weitzmann He Calls on the Mayor—Great Crowd, Carrying Zionist Flags, Gather Outside City Hall to See Noted Scientists

Thousands of members of the Jewish faith came to City Hall Park today to welcome two distinguished representatives of their race, both noted scientists and leaders in the Zionist movement, Dr. Chaim Weitzmann and Dr. Albert Einstein, who are here to promote the cause of a permanent home for Zionists in Palestine.

The principal address of welcome at the ceremonies within the City Hall was made by George W. Wickersham, ex-United States Attorney General, who was unexpectedly introduced to the delegation while Mayor Hylan, the official representative of the city, merely read a brief address.

Long before the hour set for the arrival of the Zionist leaders, hundreds of people began pouring into City Hall Park, bearded men and young women, all wearing insignia of welcome and carrying the blue and white striped flag of Zion, with the Star of David in the centre.

While Dr. Weitzmann headed the delegation, it was for Dr. Einstein, the propounder of the theory of relativity, that the throng reserved a tumultuous greeting. As Dr. Einstein left the building he was lifted to the shoulders of his colleagues in the automobile, which passed in triumphal procession through a mass of waving banners and a roar of cheering voices.

Einstein Remains Unmoved

The Mayor greeted the party in the reception room of the City Hall, where he was introduced to the members of the delegation by Bernard A. Rosenblatt. Dr. Einstein stood out in sharp contrast to the representatives of officialdom about him, accustomed as they were to all the routine of formality. He seemed to drift aimlessly through the ceremonies, with a distant, bewildered expression which caused members of the delegation to nudge him now and then into appropriate activity.

Dr. Weitzmann was introduced not only as a great scientist whose dis-

Hylan, Weizmann, and Einstein at reception at City Hall, New York.
Courtesy Corbis.

coveries had aided materially in the success of the Allied cause, but also as the leader in the Jewish struggle for a self-governing homeland. It was largely due to Dr. Weitzmann's efforts, the speaker said, that the British mandate over Palestine was secured, assuring the Jews a permanent home in the country of their origin.

"Our mission here," Dr. Weitzmann said, "is to appeal to the Jews of America, who have known so well how to serve this country both in peace and in war, to lend their helping hand in the material and spiritual rebuilding of the Jewish homeland. Thousands of the Jewish pioneers, men and women, have already made their way there in the face of difficulties and hardships, and by the toil of their hands are now engaged in laying the foundations of the New Judea."

Applause broke forth when Mr. Wickersham, in a welcoming address, recalled the fact that Dr. Einstein had held aloof from the band of German professors who issued the famous manifesto in the early stages of the war, "refusing to secede from the truth for which he stood." Likewise when he paid tribute to Dr. Weitzmann's leadership in establishing the great Hebrew University of Jerusalem, designed to be the seat of learning for those of the Jewish faith all over the world, there was prolonged applause.

Mayor Hylan dwelt upon the achievements of the Jewish soldiers in the war, and particularly those from the East Side of New York, who formed so large a part of the Seventy-seventh Division.

At the afternoon session of the Board of Aldermen a resolution was passed extending the freedom of the city to both Dr. Weitzmann and Dr. Einstein.

The New York Evening Post, April 5, p. 5.

> The throng gathered in the City Hall Park was estimated at 5,000 by the *New York Times* (April 6, pp. 1, 11), and at 10,000 by the *New York American* (April 6, p. 6) and the *New York Call* (April 6, pp. 1, 3).
>
> The chairman of the reception committee, Bernard A. Rosenblatt, remarks in his recollections that he had concluded the introduction of the guests with the following words: "Dr. Weizmann comes here on a practical mission to enlist the moral and financial support of the Jews of America in a cause which means the redemption of the prophecies contained in the Book of Books. As proof of the world in-

terest in the message which Dr. Weizmann brings, we have the privilege to welcome his associate, the world renowned scientist, Professor Albert Einstein, who comes here especially in the interest of the Hebrew University in Jerusalem."[15]

It was Rosenblatt who proposed that the City Council honor Weizmann by offering the keys of the city of New York. "As he was accompanied by Professor Albert Einstein, who was only then coming slowly into the public eye," Rosenblatt continues, "I suggested a double honor for both the two distinguished visitors."

The last paragraph of the newspaper article on the extension of freedom expressed only a wishful thought. What really took place in the Aldermen's Chamber is revealed in the next chapter.

Mayor Welcomes Jewish Mission

Dr. Chaim Weizmann, president of the World Zionist Organization, Prof. Albert Einstein, and other members of the Zionist mission which arrived here last Saturday, were received by Mayor Hylan in his office at the City Hall on Tuesday. This is said to be the first time that a mayor of New York officially welcomed Jews as Jews.

It was thought that the Board of Aldermen on Tuesday would honor Professors Einstein and Weizmann by granting them the freedom of the city. Action on this measure was stopped at the last minute by the objection of Alderman Bruce M. Falconer. This precipitated a row, during which charges of anti-Semitism were hurled at Falconer by Alderman Friedman and by City Court Justice Gustave Hartman.

After Magistrate Bernard A. Rosenblatt had introduced each of the visitors to the mayor, he addressed them as follows:

"I am very much gratified at the opportunity here afforded me to meet Dr. Weizmann, president of the World Zionist Organization, and the other members of his official party. It is my privilege on this occasion to extend to our visitors, on behalf of the city of New York, a hearty greeting and a cordial welcome.

"I understand that Dr. Weizmann comes to this country as the official representative of the World Zionist Organization, and that his purpose is to seek the complete support of all the Jews of America for the upbuilding of Palestine and creating there of a Jewish national homeland with self-governing institutions.

"As mayor of this city, the home of more than one-third of the Jews in America, I gladly join in felicitating those who have already accomplished so much toward the restoration of Palestine. The success thus far achieved may be regarded as a happy augury that continued endeavor will result in the final and complete attainment of the hope and aspiration of the Zionist organization. From an economic as well as a humane side, the mission of the Zionist organization is worthy of the fullest measure of success."

The principal address of welcome at the ceremonies within the City Hall was made by George W. Wickersham, ex-United States attorney-general, who was unexpectedly introduced to the delegation. In the course of his speech, he recalled the fact that Dr. Einstein had held aloof from the band of German professors who issued the famous manifesto in the early stages of the war, "refusing to secede from the truth for which he stood."

In responding to the generous and hearty words of welcome of all, Dr. Weizmann, as chief and spokesman of the delegation, said:

"Our mission here is to appeal to the Jews of America, who have known so well how to serve their country both in peace and in war, to lend their helping hand in the material and spiritual rebuilding of the Jewish homeland. Thousands of Jewish pioneers, men and women, have already made their way there in the face of difficulties and hardships, and by the toil of their hands are now engaged in laying the foundations of the New Judea."

The American Hebrew, April 8, p. 576.

5

Freedom of City Is Refused

✦

Tuesday, April 5

Freedom of City Is Refused to Prof. Einstein

Honor to Visiting Zionist Leaders Is Held Up at Aldermanic Meeting by Bruce Falconer

The freedom of the city, which has been extended the Prince of Wales, King Albert of Belgium, Rene Viviani among a score of others in the past few years, with [was] withheld from Dr. Albert Einstein, the world famed exponent of the theory of relativity of space, and Professor Chaim Weitzman, president of the World Zionist Organization, when the courtesy was proposed in the Board of Aldermen yesterday afternoon.

The consideration of the affair, in the board, was accompanied and followed by scenes of uproar, in which, on one occasion, heated words gave way to an exchange of blows. No arrests were made. For more than an hour after the board had officially adjourned little knots of men and women gathered around the different aldermen and held a free-for-all protest over the action.

The passage of the resolution providing the extension of the city's freedom failed because of the insistence of Alderman Bruce Falconer, who admitted that he had never heard of Dr. Einstein and had no intention of voting for the resolution until he had. Charges of anti-Semitism were hurled against him. These he denied.

Owing to Falconer's objection the resolution was forced until the next meeting. Apprised by President La Guardia of what had taken place, Mayor Hylan last night issued a call for a special meeting of the Board of Aldermen, to be held this Friday afternoon. Yesterday's melee followed an official welcome to the distinguished Zionists by Mayor Hylan himself.

In one of the most remarkable demonstrations that the city has ever witnessed, more than 10,000 men, women and children gathered in front of City Hall steps to pay tribute to the Zionist leaders. The first of the huge throng appeared at scene before 8 o'clock yesterday morning.

Only men with long beards, who seldom ever leave the confines of the Ghetto; little old women in black, with brown wigs; whole families plentifully supplied with the blue and white Zionist flags, embroidered with the Mogen Dovid, composed the greater part of the throng as the hour for the exercises drew near.

It took the greatest effort on the part of the scores of police present to prevent the crowd from surging in around the autos of the party of Zionists when it finally arrived. The steps of City Hall held the large committee of welcome appointed by the Mayor and the latter, with his official staff.

In addressing his welcome, Mayor Hylan felicitated the Jewish youth of the land, which, he said, has stood strong as a supporter of America during the past years. He said that the Jews of the East Side are to be particularly complimented for the services during the war. Following the Mayor's brief address, Colonel G. W. Wickersham spoke. He termed the scientist, of whom the member of the Board of Aldermen had no inkling of knowledge, "the greatest scientist since Copernicus."

Dr. Weitzman replied to the Mayor. In perfect English he expressed thanks for the welcome.

"Our mission here," he said, "is to appeal to the Jews of America to lend their helping hand in the material and spiritual rebuilding of the Jewish homeland. My illustrious colleague, Dr. Einstein, is here particularly on behalf of the Hebrew University at Jerusalem, where Jews may seek the higher education so frequently denied them in many lands."

Aldermen Zeltner, Graubard, Moritz and Hannoch joined in signing the resolution which was presented at the afternoon session of the aldermen. After recognizing the achievements of the visitors, it proceeds to extend the freedom of the city to Dr. Weitzman, Prof. Einstein and their colleagues.

After the reading of the resolution, Alderman Falconer, a Republican, rose with the expression of a desire to know, "who the gentlemen were" and "where they came from." Alderman Graubard asked for the floor to give the information, but only stated that Falconer ought to know who the men are if he had read the daily papers. None of the signers of the resolution ventured to explain any further. When this became evident Falconer announced that he would object to immediate consideration of the resolution.

It devolved on the Socialist aldermen to inform the board as to the exact position of the distinguished guests of the city. Alderman Vladeck explained, saying that he was not a Zionist, but nevertheless recognized the movement as a great one and deserving of respect because of its serious intentions. Dr. Weitzman was the international head of the organization and Dr. Einstein was the proponent of the theory of relativity of space.

"Prof. Weitzman," offered Alderman Beckerman, "perfected T.N.T. which, of course, Alderman Graubard knows all about. Dr. Einstein founded the theory of relativity of space, which, I am sure, Alderman Zeltner knows all about."

This did not satisfy Falconer. He had never heard the names of either, he contended. He admitted charges made by Alderman Hannoch, that he had spoken in levity because of this. Too many men have been given the freedom of the city of late, he maintained.

Falconer Stands Out

A number of aldermen and City Magistrate Gustave Hartman, who was present, tried to prevail on Falconer to withdraw his objection. They tried to impress him with the greatness of the two Zionists. It was to no avail, however. Falconer told Judge Hartman to submit a memorandum to him next week. It was protested on the floor that failure to pass the resolution would mean an insult that could not be erased by any future act.

Desperate, the proponents of the resolution succeeded in gathering the members of the board's Committee on Rules for a meeting in an anteroom. The committee came in with a report suggesting that the rules be suspended and thus permit the consideration of the resolution despite Falconer's objection. This report also needed unanimous consent for passage and, therefore, failed before Falconer, who still stood stolid.

All efforts to induce Falconer having failed, the board adjourned with the resolution on the calendar for the next meeting. The meeting ended in an uproar, at the top of which could be heard the voice of Alderman Kelleher telling Falconer that he ought to be sent to the district school to study current events.

A Doubtful Honor

Just previous to adjournment, Alderman Shiplacoff, Socialist, said:

"Alderman Falconer has the right to be wrong and foolish, if he wants to. I am sure that our noted visitors would rather forego the honor we are considering, if they knew what it is provoking. The granting of the freedom of the city is a huge joke anyway. There is no freedom in this city. Even Professor Einstein would have to look long and hard for it if we told him he could have freedom in this city."

The floor of the aldermanic chamber was crowded by those who had listened in the gallery to the discussion. They surrounded the aldermen and expressed their opinions on Falconer and his stand. Particularly irate were the other Republican aldermen. Leader Ferrand charged that Falconer had taken his stand because he wanted publicity.

Judge Hartman charged Falconer with anti-Semitism.

"You are a liar," retorted the latter. "I did not know these men."

"It is impossible that you should not know Professor Einstein," Judge Hartman replied. "It is a question of race with you."

"Oh," exclaimed Falconer, "is he the exponent of the theory of relativity. If I had known that I would have voted 'aye' on him."

"I will see to it that you are retired to private life," said Hartman. "I will fight you before the people and I will go into the councils of the Republican party to fight you."

Spectators Start Fight

Meanwhile, the arguments continued in many parts of the chamber. One of the knots witnessed an exchange of blows between two men who had formerly been spectators. They were separated and made a quick getaway, being warned of the approach of police.

Last night when the smoke had lifted around City Hall, when the announcement of the special meeting Friday had come from the Mayor's office, Justice Hartman appeared with another little sidelight on the visit of the Zionists. One of the welcomers, he said, had stolen a gold lorgnette worn by Mrs. Einstein. It is a family heirloom, he said, and a reward will be given if it is returned to his office in the court house.

The New York Call, April 6, pp. 1, 3.

The *New York Call,* a Socialist newspaper, gave more space to the Socialist aldermen's fight for Einstein's "freedom" with an alderman who was, as Rosenblatt puts it, "from the Republican 'silk stocking' district."[16]

After Magistrate Gustave Hartman had told Falconer that Einstein was "the greatest scientist since Copernicus . . . all the reply that this man Falconer could give me [Hartman] was that he would be glad if I would write out a memorandum about Einstein and Weizmann for him to read. A memorandum! My God! What do you know about that! A memorandum!"[17] "Before long the poor fellow had to retire from public life," Rosenblatt adds, perhaps with a bit of satisfaction. There is more about Falconer's objections in the next item.

Among those who were mentioned as recipients of the freedom of the City of New York, René Viviani was deputy prime minister and minister of justice of France when, as a member of a prestigious French delegation, he visited the United States in 1917.

Holds Up Freedom of City to Einstein

Alderman Falconer Blocks Move to Grant Official Honors to Two Scientists— Never Heard of His Theory—Alderman Friedman Shakes Fist in Face of Opponent and Calls Action an Insult

There is at least one man in New York who never heard of Professor Albert Einstein, whose theory of relativity has been discussed for many months in newspapers and magazines. He is Alderman Bruce M. Falconer, whose lack of acquaintance with Professor Einstein's fame caused a row in the Board of Aldermen yesterday and resulted in the freedom of the city being temporarily refused to both Professor Einstein and Professor Chaim Weizmann, chemist and inventor of the high explosive trinitrotoluol.

At the request of Aldermanic President LaGuardia, Mayor Hylan has called a special meeting of the Board for next Friday at 1:30 P.M., to take action on the resolution.

"I am expressing the feeling of the entire Board when I ask you to call this meeting in order that the desires of the people of this city may be carried out in extending this call to these distinguished people," he said to the Mayor.

Professor Weizmann is President of the International Zionist Organization, and, with Professor Einstein, M. M. Ussischkin, and Dr. Benzion Mossinson, is here to confer with American Zionists. They were received at the City Hall yesterday by Mayor Hylan, and a committee of citizens. More than 5,000 Zionists filled the plaza in front of the City Hall.

It was thought that the granting of the freedom of the city to the two visitors would be a mere formality. So it would have been but for Alderman Falconer, who is a lawyer and lives at 701 Madison Avenue. After the ceremony the Aldermen went to their Chamber and a resolution was introduced by Aldermen Louis Zeltner, Moritz Graubard and Samuel R. Morris in honor of the visitors. Every one was ready to vote favorably when Alderman Falconer arose. He confessed that until yesterday he never had heard of either Professor Einstein or Professor Weizmann. He asked to be enlightened, but nobody offered to explain the theory of relativity. Mr. Falconer said that he thought the freedom of the city had been too often granted, and, although his objection had nothing to do with racial or religious prejudices, he believed that caution should be exercised.

A storm broke about Alderman Falconer's head. Laughter and protests came from every side, and several members tried to tell him the records of the two men, but their recital made little impression upon the Alderman.

Rules Committee Dodges

A motion that the resolution be made a general order for next week when it could be passed over Alderman Falconer's protest precipitated a parliamentary row, and in a few minutes the board was tangled up in rulings. President LaGuardia came in and took the chair. He ruled that the point of order to make the resolution a general order was debatable, and about this time the Committee on Rules, led by Alderman Kenneally, slipped out of the room.

Alderman Falconer was obdurate, and at the end of the debate the Rules Committee came back and an attempt was made to get around his objection. It was moved to suspend the rules, when the resolution could be passed over his objection. But Alderman Falconer suspected the purpose of the motion, and objected. Alderman Friedman then asked that the resolution be withdrawn.

After the incident was officially closed there were angry arguments in the boardroom. Alderman Friedman shook his fist under Mr. Falconer's nose and said that his action was an insult and that it would carry the issue into

Mr. Falconer's district. Judge Gustave Hartmann tried unsuccessfully to tell Mr. Falconer what Professor Einstein had done in science.

After the adjournment of the meeting Judge Hartman charged Alderman Falconer with having made his objection to the resolution because of purely anti-Semitic motives. This brought a denial from the Alderman and when Judge Hartman repeated his charge Mr. Falconer said: "You're a liar, I am most certainly not opposed to the Jewish people as a race."

"I will not let this matter drop," said Judge Hartman. "Not only will I bring the matter before the people of the city and the intelligent Jewry, but I will also press this matter in the council of the Republican Party. I am firmly convinced that your attitude in this matter was prompted by anti-Semitism, and I will not be satisfied until you are retired from public life."

When Professors Weizmann and Einstein arrived at the City Hall, accompanied by their wives and other members of the delegation, they were escorted to the Mayor's office by James F. Sinnott, Secretary to Mayor Hylan, and the Committee of Welcome led by Magistrate Rosenblatt.

"As Mayor of this city, which is the home of more than one-third of all the Jews in America," said Mayor Hylan, "I gladly join in felicitating those who have already accomplished so much toward the restoration of Palestine. The success thus far achieved may be regarded as a happy augury that continued endeavor will result in the final and complete attainment of the hope and aspiration of the Zionist organization.

"May I say to Dr. Weizmann and Professor Einstein that in New York we point with pride to the courage and fidelity of our Jewish population, demonstrated so unmistakably in the World War."

George W. Wickersham, former Attorney General, also spoke of the achievements of the two leaders of the delegation.

Professor Weizmann thanked the Mayor and Mr. Wickersham for their welcome, which he accepted as showing sympathy for the cause he represented.

Mrs. Einstein lost a gold lorgnette with a chain attached during the reception at the City Hall. It was an heirloom.

The New York Times, April 6, pp. 1, 11.

6

Freedom of the City Is Given

✦

Friday, April 8

Zionist Leaders Given Freedom of N.Y. State

At the instance of the Hon. Nathan Straus, Jr., State Senator, the New York State Legislature has conferred the "Freedom of the State of New York," upon Dr. Weizmann and Prof. Einstein.

The text of the resolution which was unanimously adopted is as follows:—

"Whereas Albert Einstein of Switzerland and Chaim Weizmann of Great Britain are now visiting our State; and

"Whereas the purpose of their visit is to cement the bonds of unity between the United States and her neighbors abroad in the great struggle for human progress and happiness, and especially to unite the old world and the new in establishing a cultural centre for the Jews of the world in Palestine; and

"Whereas the achievements of Dr. Einstein in the spheres of physics and astronomy have commanded the attention and the admiration of the entire civilized world, and the record of Dr. Weizmann as a chemist during the World War has made the people of the allied and associated powers his debtors, and,

"Whereas it is the desire of the Commonwealth of New York to make these distinguished visitors feel that every true American heart goes out to them in cordial welcome; therefore,

"Be it resolved that the people of the State of New York extend to Dr. Einstein and Dr. Chaim Weizmann and their associates the handclasp of fellowship and a heartfelt welcome."

Yidishes Togeblat—The Jewish Daily News, April 8.

Einstein Given Freedom of City by 57 Votes to 1

Falconer, Standing on Own Brand of Americanism, Still Dissents from Bestowing Courtesy

Alderman Falconer still dissenting, the Board of Aldermen yesterday extended the freedom of the city to Dr. Albert Einstein and Professor Chaim Weitzman, famous scientists and Zionists, now visiting here.

In impassioned speeches several of the members attacked the alderman who had blocked the resolution at its original presentation. Alderman Collins, Democratic leader, called him a "lackey," a "bigot" and a "disgrace to the city."

Alderman B. C. Vladeck, Socialist, compared him with Newton's dog, who was unable to comprehend his master's theory of gravitation; the Republican leader, Alderman Ferrand, disclaimed all responsibility for Falconer's stand on the question.

As his defense, Falconer offered his contention that too many men have been given the freedom of the city. Einstein, he said, is a German, and to honor him would be to honor an "enemy alien." Weitzman, he admitted, might be well known to Jews, but is not sufficiently distinguished to warrant the respect of the Board of Aldermen.

He Was a Pacifist

"With all of Falconer's professed Americanism," said Vladeck, "Professor Einstein is a better citizen than he is. It is well known that Einstein was a pacifist and refused to join in Germany's war; that less than five months ago he was mobbed on the streets of Berlin by a gang of reactionary students, who tore his scalp and broke his glasses."

When the discussion had ended and a roll call was demanded, Falconer, despite the information offered by Vladeck, voted against the extension of the courtesy. The vote was 57 to 1. A committee of eight was elected to visit Einstein and Weitzman on Monday.

Reading from his speech, Alderman Falconer recounted his family history. He had an ancestor, he said who was French, became naturalized as an Englishman and later came to America as the secretary to Lord Cornbury. The latter, he continued, was the first man who ever received the freedom of New York City.

"The granting of the freedom of the city is the greatest honor that can be bestowed on any visitor to this city," said Falconer. "There have been too many given that in the past two years. This country must not become a forum for the airing of foreign political questions. America for Americans. America first."

Alderman Collins ridiculed Falconer's plea of Americanism. He sharply criticised him for his stand against extending the freedom of the city to Archbishop Mannix and Mrs. MacSwiney when they were in the city.

An Old-Style Secretary

Referring to Falconer's ancestor, Collins said:

"In the old days, when Lord Cornbury came here, there were no typewriters. There was no system of correspondence. A secretary did not write or read letters. He was a servant. He shined shoes. He was a lackey, that is, he was a Falconer. It was narrowness, bigotry and a disgrace for Falconer to take the stand he did."

Discussing the resolution Alderman Vladeck said:

"It is men of Falconer's type who consider that those who disagree with them are un-American. Great men, such as Debs and Lincoln, have suffered from such. What are we to think? Falconer says 'America First.' Shall it be his America of persecution or shall it be an America of broadminded men and women?

"Falconer's 'America First' means the same thing as Wilhelm the Second's 'Germany First.' He meant not the Germany of culture, of science, but the Germany of the mailed fist. Falconer's sentiments are the same.

"Falconer is a reactionary in politics just as there are reactionaries in science. He does not recognize Einstein's theory of the relativity of space. Newton, who discovered the law of gravitation had as a companion, a little dog named Diamond. Newton discovered the law by observing an apple fall downward. Diamond saw the apple fall, but never evolved or recognized the theory of gravitation.

Falconer's Clothes and Shoes

"Falconer claims that the 250 years his family has spent in this country makes him a superior American. It has not. It has made him an American who does nothing. The bridge which a few feet from here starts to span the wide river was built by Irishmen, by Italians, Bohemians, Lithuanians, Germans and Jews and other foreigners.

"Falconer's house was built by Sullivans, McCarthys and MacSweeneys; his clothes were made by Cohns and Goldsteins; his shoes were made by Tony and Joe. Falconer's only claim to existence is the past, not the present. In 50 years there will be no room for Falconers, the foreigners will watch out for that."

He has fought the Zionist movement for 15 years, the Socialist said, and would continue to fight it. He believed that it could not and ought not succeed. Nevertheless, he said, it is a cause in which many people are seriously interested and is deserving of respect as such.

The New York Call, April 9, pp. 1, 6.

Although at the height of the quarrel it sounded great, neither Einstein's scalp nor his glasses were at risk in Berlin. It is true that on February 12, 1920, when Einstein was lecturing at the University of Berlin, students protested against nonstudents having been admitted to listen and occupy seats in the auditorium. The newspapers even mentioned some anti-Semitic remarks, but Einstein at the time denied that the protest was against him as a Jew.[18]

On Mannix and Mrs. MacSwiney, the other *personae non gratae* for Falconer, see p. 64.

To mention the Socialist union leader Eugene Victor Debs in the same breath with Abraham Lincoln was a strategy of Debs's supporters at the time. Sitting in jail for campaigning for U.S. neutrality in the war, he even ran for the presidency. He was released from prison by President Harding.

Freedom of the City Is Given to Einstein

Alderman Falconer Casts Lone Negative Vote at Special Meeting of the Board

Freedom of the City of New York was granted to Professor Albert Einstein, discoverer of the theory of relativity of space and time, and Dr. Chaim Weizmann, president of the World Zionist Commission, by the Board of Aldermen at a special meeting yesterday. The vote was 56 to 1.

Alderman Bruce M. Falconer, who on Tuesday prevented the resolution being passed by refusing unanimous consent, cast the lone negative vote yesterday.

William T. Collins, Democratic leader, led a verbal attack on Alderman Falconer for his opposition, backed by August Ferrand, Republican, and B. C. Vladeck, Socialist leader. Alderman Thomas M. Farley, Democrat, told Falconer he was trying to create class hatred in the board and in the city.

Narrowness and Bigotry

Said Mr. Collins:

"It was only narrowness and bigotry that made the one member of this board object to granting the freedom of the city to Professor Einstein and Dr. Weizmann.

"Alderman Falconer is a descendant of a man who came to this country in 1702 as secretary to Lord Cornbury. In those days the job of the secretary was to blacken the shoes of his lord. Those people opposed Washington, and there is no doubt in my mind that a man with such an ancestry would oppose any Irishman or any Jew. He ought to give up America and go back to England."

Alderman Falconer, replying, said:

"I stand for the principle of America first—first in war, first in peace, and first in the hearts of its countrymen. My objection is based only on this principle. Alderman Friedman did me a great injustice Tuesday when he said I objected because of prejudice against the Jewish race. Some of my best and closest friends are Jews!"

To Hold Reception

New York Jewry will give a reception to the Zionist commission Tuesday night in Sixty-nine Regiment Armory. All Jewish elements will participate.

Representatives of more than 1,500 Jewish societies form the committee, of which Judge Gustav Hartman is chairman. Preparations are being made to seat wounded World War veterans. Jewish legionaries, who fought in Palestine under General Allenby, will escort them to the armory. Dr. Nicholas Murray Butler, president of Columbia University, will speak.

Professor Einstein yesterday visited the College of the City of New York.

The New York American, April 9, p. 5.

It is difficult to determine whether there were 56 or 57 Aldermen for Einstein; more important is, however, that there was only one against him. I don't understand why this resolution passed now, for Falconer did not change his mind. And if it passed, why had it not passed earlier?

Freedom of City Given to Einstein

Aldermen Honor Relativity Discoverer and Prof. Weizmann Despite Falconer's Protest—He Defends Adverse Vote—Cites Courtesies to Dr. Cook, De Valera, Mannix and Mrs. MacSwiney as Mistakes

Professor Albert Einstein, the noted mathematician and discoverer of relativity, and Professor Chaim Weizmann, British chemist now have the freedom of New York City. It was voted to them yesterday at a special meeting of the Board of Aldermen, made necessary by the refusal of Alderman Bruce Falconer to consent to the passage of the resolution when it first came up on Tuesday, when the two scientists were welcomed by Mayor Hylan at City Hall.

Alderman Falconer cast the only negative vote yesterday, and in so doing said he was not actuated by race prejudice but that he had in mind the dignity of the honor which has been given to some of the greatest Americans, and thought it should not be conferred on any one unless he were known to every person in the city. He said his first ancestor in this country came as secretary to Lord Cornbury, the first person to receive the freedom of the city, in 1702.

Alderman William T. Collins, leader of the Democrats, seized upon the mention of Alderman Falconer's ancestors with avidity, and ridiculed it.

"We on this board are just as proud of our city and of the conferring of the freedom of the city on guests as is Alderman Falconer," he said. "It was only narrowness and bigotry that made the one member of this board object to granting the freedom of the city to Dr. Weizmann and Professor Einstein."

Alderman Falconer said that Alderman Friedman did him a great injustice in saying that his objection was based on race prejudice, and said that his private physician is a Jew and that many of his friends are Jews.

"In 1909," he said, "the keys of the city were unfortunately given by the Board of Aldermen to Dr. Cook, who pretended to have discovered the North Pole, but were afterward officially withdrawn from him. After that the freedom of the city was not again extended for ten years, until the second year of the Hylan Administration, when it was given to Eamon de Valera, at a meeting which occurred when I happened to be away from the city.

"Since that time it has been extended to Cardinal Mercier, King Albert of Belgium, the Prince of Wales, Archbishop Mannix and Mrs. MacSwiney. At the time the resolution was suddenly proposed in connection with Archbishop Mannix, I did not vote in favor of conferring the honor upon him.

"The next and last individual upon whom this honor was conferred was Mrs. MacSwiney. I did not vote for it, and if I had had a proper chance would have objected.

"I have been assured," he said, "that Professor Einstein was born in Germany and was taken to Switzerland, but returned to Germany prior to the war. He is consequently a citizen of Germany, of an enemy country, and might be regarded as an alien enemy."

Alderman Friedman told Alderman Falconer that Professor Einstein was not a citizen of Germany, but of Switzerland, and Alderman Vladeck, leader of the Socialists, also said that Professor Einstein was far from being a German citizen.

Alderman Ferrand, the Republican leader, in moving the question, said:

"For what has occurred I make no apology to this board or to the citizens of the city. It can be charged to no party. It can only be charged to an individual who is arrogant and ignorant. We will have to take it from whence it comes."

Professor Einstein visited the College of the City of New York yesterday, and attended a class in mathematics and physics, where he listened to an explanation of his theory by Prof. Edward Kasner of Columbia University. President Sidney Mezes, of the City College, and a number of advanced students were present. Prof. Einstein, who understands English, although he does not speak it well, complimented Prof. Kasner on his presentation of the subject, and later made a twenty minute talk.

It was announced at Princeton University yesterday that Professor Einstein would be the guest of the University from May 9 to 15 and would give five lectures in that time on relativity.

The New York Times, April 9, p. 11.

Let us look at the people Falconer regretted having been given the keys to New York City.

Eamon de Valera was an Irish statesman, born in New York. He was later elected a member of the British Parliament and president of the Irish separatist party Sinn Féin. Arrested in May 1918, he escaped from prison in 1919 and came over to the United States to raise funds for the Irish cause. In 1959 he was elected president of Ireland.

Archbishop Daniel Mannix was an Australian but was born and raised in Ireland; he was an ardent Irish patriot. While he crossed the United States in 1920 en route to visit the pope in Rome, he strongly spoke out against British policy in Ireland.

And then there's Mrs. MacSwiney, née Muriel Murphy, wife of Terence MacSwiney, Mayor of Cork, Ireland, a leader of the Irish Republican Army (IRA), who in 1920 had died of a hunger strike in prison.

Thus Alderman Collins was right in the previous article, when he guessed that "a man with such an ancestry would oppose any Irishman."

Désiré Mercier, Cardinal of Belgium, was the leader of the Belgian spiritual resistance to the German occupying forces during World War I. He came to the United States in 1919 as a deeply honored guest and was invited by numerous organizations and personalities, among them President Wilson. His visit was an opportunity for the American public to revive the memories of the sufferings of Belgium, and it is no wonder that in Princeton Einstein was also praised for his stance against the invasion of Belgium by the German army (see p. 166).

That Einstein was a Swiss citizen nobody doubted. Falconer was, however, not far from the truth when he suspected that Einstein must have been a German citizen. As a state official, Einstein took the oath of allegiance to the imperial constitution on July 1, 1920, and to the Prussian constitution on March 15, 1921,[19] but Einstein did not feel that these acts entailed an acceptance of German or Prussian citizenship. The question was raised sharply in November 1922 when he was awarded the Nobel Prize. He could not attend the ceremony because he was on a trip to Japan. Who would represent him—the German or the Swiss ambassador? The German ambassador turned to the German Foreign Office, the Office turned to the Prussian Academy, the

Academy turned to Einstein. It took eight months until the question was settled by declaring that when Einstein accepted the Berlin appointment to the Prussian Academy in 1913, he had implicitly accepted Prussian citizenship, without giving up his Swiss citizenship. In 1921, he could still enjoy this ambiguity (see p. 97).

For more on Einstein's twenty-minute talk at Kasner's lecture, see Morris R. Cohen's review on pp. 116–122.

The resolution of the Board of Aldermen of the City of New York is dated April 16.

> In the Board of Aldermen
> Resolution
> granting the Freedom of the City of New York to Professor Chaim Weisman, President of the World Zionists, Professor Albert Einstein and Associates.
> By Aldermen Graubard, Hannoch Morris and Zeltner.
> Resolved
> That we, the members of the Board of Aldermen, cognizant of the arrival of Professor Chaim Weismann, President of the World Zionist Organization, accompanied by Professor Albert Einstein, one of the most distinguished scientists of our age, and appreciating the noble service that they have rendered the Jewish people in their valiant struggle to realize their age-long aspiration of re-establishing a national home in Palestine, hereby sincerely and cordially tender unto these honored guests the freedom of the City of New York, and urge our fellow citizens to join us in these, our heartfelt felicitations.
> Adopted by the Board of Aldermen April 8, 1921. A majority of all the members elected voting in favor thereof.
>
> Approved by the Mayor April 16, 1921.
> M. J. Cruise
> Clerk[20]

Curiously enough, Einstein received the key of New York City a second time in 1931. Perhaps he left the first one in Berlin. . . . But that's another story.

One of the reasons for the invitation to the College of the City of New York must have been to show him how the Americanization of Eastern European Jews had occurred there. "For many Eastern European Jews, going to City College became synonymous with escape from Lower East Side poverty," Allon Schoener writes. "In the 1920s and 1930s, they accounted for four-fifths of the student body, which reached a high point of 32,030 in 1929."[21]

An article in the *Jewish Independent* weekly newspaper (April 15, p. 1) is a compilation of the articles published on April 9; therefore I omit it here.

Einstein Finds Man Explaining Relativity

Discovery of One Who Understands Theory Followed by "Freedom" of New York

NEW YORK, April 8.—Prof Albert Einstein's luck improved today.

First he discovered a man in New York who understood the Einstein theory. Then he had conferred on him the freedom of the city, which was denied him earlier in the week.

The visitor dropped in this morning on a class in mathematics and physics in the College of the City of New York. Prof. Edward Kasner, of Columbia University, was expounding the Einstein theory of relativity. When Prof. Kasner had finished, Prof. Einstein complimented him and then cautioned the students not to keep their noses too close to the grindstone if they sought success in science.

The Public Ledger—Philadelphia, April 9, p. 3.

Einstein Finds America "Solid"

Thinks Scientists Here Have Forged Far Ahead—Professor Who Propounded Theory of Relativity Has Other Things to Talk About

Prof. Albert Einstein sat in his room on the seventh floor of the Hotel Commodore and heaved a care-worn sigh. Another interview. Wearisome things, he seemed to be thinking, when they come in such a perpetual procession. But Prof. Einstein is nothing if not polite, and he sat waiting in stoical silence for the inquisition to begin, absently twisting the stem of his glossy brier pipe. His handclasp had been firm and friendly, and there was an unconscious dignity about this man of medium height and build that commanded instant respect.

One could hardly allude to the lineaments of this Swiss Jew as Semitic. They are hard to classify, racially—and still more difficult professionally. Snap judgment would place him as an artist, perhaps a musician, and queerly enough such is the case; not so strange, after all, when one remembers that Copernicus was also a painter. But his violin is an avocation, perhaps a recreation, with this physicist who has had mental adventures in grappling with time and space that dwarf the physical excursions of men of lesser imagination.

It takes inconceivable intellectual courage to leave all conventional concepts behind and embark upon the un-known, the unconceived, as Prof. Einstein has done, and that courage glowed from his calm dark eyes as he glanced out of the window at the April sunshine which poured down on a nearby building.

And although they were thoughtful, there were kindly wrinkles about the corners, fugitive crow's-feet but spontaneous, ready at a second's notice to lend to that dreamy artist's face a humorous cast. Long hair fluffed up into a curly aureole, black turning to gray, crowned the head.

American Solidarity Reassuring

But Prof. Einstein was manifestly unhappy, and he caressed his pipe more vigorously than ever. Interviews are odious things anyway, when the interlocutor speaks in a foreign tongue, and if the opinion of the latter may be interpolated it is equally distressing to him to resort to an interpreter; there had been a time when he had plowed through dusty scientific articles in *Die Welt der Technic,* and had been able to twist from his Anglicized larynx an agonized umlaut, but since that time there had been a war, when a throaty vowel attracted suspicion, and other languages had become more popular, and now he was able only to smile at the Professor and address his questions to the interpreter. But further reference to this clumsy machinery must be omitted

"My impression of the American people is by necessity a limited one," said Prof. Einstein, waving his hands in a deprecatory gesture as he interjected the statement that he had not yet had time to draw proper conclusions. "But the suggestion I have got so far is that of solidity, of soundness. In Europe so many people seem to have lost control of their mental governance, but here there are a solidity and a solidarity that are reassuring."

"What do you think of New York's architecture?"

He smiled, and resorted to relativity, not spatial, but terrestrial, saying in his soft urbane voice:

"I can't say. I'm at a loss to form any opinion yet of your structures such as the Woolworth Building. You see, I have nothing to compare it to. I have never seen anything remotely like it before."

"And American physicists, astronomers, scientists? Do you think they are as advanced as those of Great Britain and Europe."

"I believe they have forged ahead," was his surprising reply. "During the long period of war the European scientists were distracted from their work, and perhaps this is the cardinal reason that your men have surpassed them. American scientists have done, and are doing, excellent work."

He was asked what he thought of American universities.

"I'm hardly able to say"—he paused, and poured out to the interpreter a flood of smooth syllables, from which the interviewer was able to glean that Prof. Einstein thought it perhaps presumptuous to pass judgment where he had so little firsthand information. "But," he continued, "I may say that I think highly of them, indeed, and admire their progressive spirit. One of their outstanding merits is the way they have induced foreign scientists to come to them and the way they adopted them and made them a part of themselves."

During the questioning the stem of the treasured brier had been twisting with increasing rapidity, and at this juncture Prof. Einstein informed his listener good-naturedly but firmly that he was visiting America in only one capacity, as a member of the Zionist World Organization, and it was his desire to talk on this subject alone.

For University at Jerusalem

"I am lending my whole energies at present," he said, "to the founding of a university at Jerusalem. The work was begun even during the war. A building, far insufficient in size for our ultimate purposes, has been obtained and an incomplete library secured. But the bulk of the work remains to be done.

"The university is necessary for two reasons. First, to establish a cultural and spiritual centre for the Jewish people in Jerusalem, and, second, because the need of such an institution is keenly felt by the younger people of Palestine. All over the world the Jewry are contributing their efforts and money. Currency has so depreciated in most countries, with the exception of England and America, that their contributions can be of no great moment, but it is the people of these countries who are giving their actual physical aid, and to the Jewry of England and America we must look for the other half of the collective contribution, the financial half. With the present unsettled world economic conditions we look to the half million Jews in England and the three million in America for funds, and to those of the other countries for the physical labor, the actual building and breaking of ground."

A special fund has been created, he said, separate from the general fund of the Zionist World Organization, for the purpose of setting up this university. Prof. Einstein's efforts will not cease with the completion of this university. He will continue his connection, he said, perhaps in the capacity of a visiting lecturer, or he may even occupy a permanent chair.

He laid his pipe aside. The interview was seemingly at an end. But the interviewer, with a desperate feeling that this was not just the kind of report one should get from the propounder of the theory of relativity, made a final plea to be allowed to put some questions of at least a semiscientific nature. Prof. Einstein is too gracious to be stern. He weakened. He relented. Just one question. No more.

And from his answer to this question one may draw the conclusion that those people of large girth who desire to decrease their dimensions need only to acquire a high velocity, and as that velocity increases, their dimensions will decrease. But, of course, the Interborough achieved that end— during rush hours—long ago, enforcing reduction of dimensions independent of velocity, although it may be held forth as an extenuating circumstance according to the theory of relativity that the trains grow shorter as they hurtle through the subways.

But this is a wholly jocular deduction, for the question put to Prof. Einstein was:

"Do dimensions of moving objects on earth with respect to a stationary point slightly change at comparatively slow velocities? For example, does a train moving past an observer at a rate of one mile a minute have smaller dimensions, relative to the observer, than it would have when at rest?"

"There would be a change, but it would be infinitesimally small," he said, amused at the question, and he wrote the formula which indicated the ratio of change—the square of the velocity in kilometres per second of the moving object, divided by twice the square of the velocity of light in kilometres per second. And inasmuch as the velocity of light is 300,000 kilometres a second, the interviewer agreed that no one would worry about the change so far as subway trains were concerned.

The New York Evening Post, April 8, p. 7.

The introductory paragraphs of a long article in the *American Hebrew,* reprinted below, reveal that it was not only America that amazed Einstein but also Einstein who amazed America.

Prof. Einstein—"The Poet in Science"

Although in this country less than a week, Prof. Albert Einstein, Jew, "poet in science," "intuitive physicist," propounder of the theory of relativity and missionary for the Hebrew University of Jerusalem, "has come, has seen, has conquered."

The American press has taken him for its very own. His briar pipe and his theories regarding natural laws playfully are mingled in newspaper paragraphs. A description of his temperament and violin finds itself sandwiched between statements concerning the Zionist cause.

In the party with Prof. Einstein aboard the steamship Rotterdam, which docked last week, were Prof. Chaim Weizmann, president of the Zionist World Organization, discoverer of trinitrotoluol, and head of the British Admiralty laboratories during the war; Michael Ussichkin, a member of the Zionist delegation to the Paris Peace Conference and now Resident Chairman of the Zionist Commission in Palestine, and Dr. Ben-Zion Mossensohn, president of the Hebrew Teachers' Organization in Palestine and principal of the Jaffa Gymnasium; Gerson Agronsky, head of the Zionist Commission Press Bureau; Solomon Ginsburg, son of the famous Jewish philosopher Asher Ginsburg (Achad Ha'am), and L. Stein, of the London Zionist Office.

Discusses His Theory

Among the fervent and orthodox in the ranks of the Zionists in this country the reaction of the American public and press to the visiting delegates is understood to have produced something akin to a profound shock. There is reason to believe that Prof. Einstein was induced to accompany the mission in the hope that his presence would act as a "tail to the kite." And now, lo and behold, contrary to all calculations (except, perhaps, those based on the theory of relativity), the tail has become the whole kite.

[The rest of the article was cobbled together from earlier interviews already presented here.]

The American Hebrew, April 8, p. 567.

The events of the week are summarized by the *New Palestine* in the following item.

Weizmann, Einstein and Others of Delegation Get Big Welcome

New York Jewry has been expressing the wildest enthusiasm since Saturday when Dr. and Mrs. Chaim Weizmann, Professor and Mrs. Albert Einstein, M. M. Ussischkin, Dr. Ben Zion Mossensohn, Gerson Agronsky and L. Stein arrived in this country. Over 10,000 people were at the dock to see them land, and so great was the joy of welcome that they followed the automobile that took the group to the Hotel Commodore, and refused to disperse until far into the night.

When the revenue cutter which Mayor Hylan assigned for their use took them off the boat the air rang with the cheers of the crowd. The band struck up "Hatikvah" and the "Star-Spangled Banner," and after a few words of welcome by the Mayor's Reception Committee and by Judge Mack, a parade started which marched through lower New York to the Commodore, where they are staying.

All this week Dr. Weizmann, Professor Einstein and the other members of the delegations are being entertained privately. The first official public reception will be held at the Metropolitan Opera House on Sunday, April 10th. Among the speakers will be Dr. Weizmann, Professor Einstein, Mr. Ussischkin, and Judge Mack, who will be chairman. There will be a number of organ pieces as well as mass singing.

Mayor Hylan of New York City expressed a cordial welcome to the delegation at a public meeting at the City Hall on Tuesday, April 5th. After an impressive entrance to the City Hall, with 10,000 people cheering and dozens of cameras cranking, the Mayor, Attorney General Wickersham, and Judge Rosenblatt made addresses of welcome. In acknowledging their cordial greeting to himself, Professor Einstein and the other members of the group, Dr. Weizmann said that with the co-operation of American Jewry, the task of rebuilding Palestine, which fired the imagination of every idealist, would readily be accomplished.

The New Palestine, April 8, p. 1.

7

Fervid Reception

✦

Sunday, April 10

The Welcome Meeting

The great meeting of welcome to Dr. Chaim Weizmann, Prof. Albert Einstein, Menachem Mendel Ussischkin and their associates has been held and is now a thing of the past.

The big Metropolitan Opera House on Broadway and Thirty-ninth Street was crowded to capacity. Every seat was occupied and in addition the space allotted to the standees was full too. And on the stage there were hundreds of persons. It was a thoroughly representative gathering. From all parts of the country Jews had come to hear what the leaders of world Zionism had to say. And for the nonce Zionists and non-Zionists sat cheek by jowl.

The meeting was called for two thirty. It was fully an hour later that it was called to order by Judge Julian W. Mack, President of the Zionist Organization of America. All kinds of rumors were rife. Dr. Weizmann and his colleagues had been in session with the Executive of the Zionist Organization Saturday night and Sunday up till a short time before the gathering was scheduled to begin. It was known that there were sharp differences between the visitors abroad and the local leaders. Reports of resignations and splits were in the air, and some were of the opinion that Dr. Weizmann would not come to the meeting. So it was with a sigh of relief that about twenty minutes after three, Nathan Straus led the visitors and their escorts upon the stage. Mr. Straus was followed by Judge Mack, Dr. and Mrs. Weizmann, Prof. and Mrs. Einstein, Ussischkin, Dr. Schmaryahu Levin and Dr.

Benzion Mossensohn, Judge Irving Lehman, chairman of the American Branch of the Palestine Economic Council, Louis Marshall and members of the executive of the Zionist Organization of America.

With a roar such as the Metropolitan Opera House, accustomed to ovations and to scenes of enthusiasm had never witnessed the vast audience greeted the leaders. Waving flags and again and again singing the Hatikvah, they shouted for Weizmann, for Ussischkin and for Einstein. There could be no doubt of the intent of the throng. It was there to pledge them the support of American Jewry as Judge Lehman subsequently said, in the mighty task of rejuvenating Eretz Yisroel. Judge Mack in his opening speech made no reference to the conferences that had taken place between the Zionist Organization and Dr. Weizmann and his colleagues. He spoke heartily. Judge Lehman who followed, laid stress upon the fact that American Jews would never forget that they are Jews and he assured the world-Zionist leaders of American Israel's support. Mossensohn who came next dwelled upon the influence of Palestine and Rabbi Meyer Berlin, the Mizrachi leader, brought the greetings of his part in Zionism. Ussischkin, sharp and virile, pointed to the development of a complete Jewish life in Eretz Yisroel and Rabbi Abba Hillel Silver of Cleveland with a declaration that the young Jews of America would go to Palestine. There was a roar of approval. Schmaryahu Levine, who spoke in somewhat cryptic terms, wanted no compliments. He wanted American Jews to emancipate themselves and to come to Zion, instead of being begged to go. Louis Marshall did not declare himself a Zionist but he shouted that none could remain indifferent now. He made a most powerful plea for unity. "Woe to those who bring strife where peace should reign. Woe to those who bring discord where unity should exist," he exclaimed and his sentiment was cheered to the echo.

And then came what everybody was waiting for, the speech of Dr. Weizmann. Again there were cheers which lasted for several minutes.

Weizmann is no orator, but he knows what he wants to say and says it calmly and straight from the shoulder. He said nothing of the struggle he has with the American Zionist leaders. He devoted the major part of his address which lasted about three quarters of an hour, to stating what needed to be done in Palestine and he closed with what the Chalutzim were doing. He spoke under a great strain. There was no doubt that the conferences which he has had with Zionists here had left their impress upon him. He appealed for help, for the means to carry on the work.

He pleaded for the honorable fulfillment of the duty devolving upon all Jews.

And with the singing of Hatikvah the great meeting came to an end.

I. L. Bril

Yidishes Togeblat—The Jewish Daily News, April 12.

Fervid Reception to Zionist Leaders

Metropolitan Opera House Jammed in Honor of Weizmann and Einstein—Persecution Spur to Hope—Rabbi Silver, Judge Julian Mack, Judge Lehman Tell Visitors American Jews Will Aid

Every seat in the Metropolitan Opera House, from the pit to the last row under the roof, was filled, and hundreds stood at a reception given by American Zionists yesterday to Dr. Chaim Weizmann, head of the International Zionist Organization, Professor Albert Einstein, exponent of the theory of relativity, and their associates, who are in this country to obtain support for the Zionist movement. They applauded long and frequently at references made to the new hope of a home for Jews in Palestine under British protection.

"Our Jews of America will respond," Rabbi A. H. Silver told them. "Now that the world has been convinced Jewry will be convinced. And I tell you what will help and what will stimulate this mighty conviction in the souls of our people—just that new tide of anti-Semitism that has encroached upon our shores. That is not going to break our spirit and demoralize the ranks of Israel. It will strengthen and crystallize the souls of our people.

"The Jew does not yield the real oil of his spirit until he is persecuted, until he is oppressed, until the people about him show their resentment and their hatred. We are going to respond to every attack upon our people, to every libel and every slander, by more Jewishness, by more schools and synagogues and by a more intensive and loyal work in Palestine."

Rabbi Silver hailed Dr. Weizmann as the "Ezra of the second restoration, the man who is to announce the gathering anew of the scattered remnants of our people to a new and glorified life in Palestine." Professor Einstein he called an "Intellectual Titan." M. M. Ussischkin, head of the Zionist Commission in Palestine, and Dr. Ben Zion Mossensohn also were welcomed by Rabbi Silver and Judge Julian E. Mack.

"This is a solemn moment in the history of the Jewish people," Judge Mack said. "This day will remain forever in our memories."

"The American Jew has heard the call of his suffering brothers and the American Jews will answer that call from their own plenty," Judge Irving Lehman said. "We must do all we can to relieve their needs. The work of relieving their physical distress must not slacken. Much has been given; still, few have given in accordance with their means or in accordance with the needs of the hour. Our enemies have thrown at us the daunt of the Jewish solidarity. Let us show to ourselves and the world that the Jewish solidarity is not lacking when the call of duty is sounded."

Louis Marshall, who was a delegate to the Peace Conference, representing American Jews, also welcomed the visitors.

"Whatever our views have been in the past," he said, "no Jew who loves his people and is proud of their history can remain indifferent now. The great nations of the world have spoken. The opportunity has come for the Jews to realize their dreams. Palestine can be made again the home of prophetism. It was from there that the fiery words of prophecy went forth and the world learned the meaning of justice, love and mercy. All the Jews should help to build up Palestine, not only as a home for homeless Jews, but as a religious, a spiritual and a cultural center. You cannot build up a land by enthusiasm. Action is necessary, and liberality, generosity and sacrifice are necessary; and above all unity in Israel is necessary. I feel that in all these the Jews of America will not be wanting."

Dr. Weizmann asked the audience to stand as a tribute of appreciation for what the United States has done for the fulfillment of Jewish aspirations in Palestine, and said there was no doubt Jews could develop Palestine agriculturally and industrially if they get the opportunity and support.

"I know many will question whether the Jew will be able to do hard work," he said. "The answer has been given by the numbers who have poured into Palestine in the last nine or ten months. You can see the young men at work today, fine, upright, clean, strong, young men, singing to the tune of the hammer on the stones as they work on the roads of Palestine. They find hard work, but ask only if it will last and go cheerfully at it."

The New York Times, April 11, p. 10.

"Welcome, builders of the Jewish home!"
Yidishes Togeblat—The Jewish Daily News, April 3, p. 1.

3,000 Zionists Give Weizmann Great Ovation

Jewry's Most Noted Americans Welcome Leaders of Movement Seeking Aid for Palestine University

Three thousand enthusiastic Jewish men and women gathered in the Metropolitan Opera House yesterday afternoon and with songs, cheers and prolonged applause welcomed the first appearance here in public of Prof. Chaim Weizmann, president of the World Zionist Organization.

One after the other numerous speakers stirred the audience to great applause with eulogies of Professor Weizmann and Dr. Albert Einstein, famed scientist and Zionist, who sat on the platform. The occasion was marked by the attendance of many Zionists and Jews of national reputation.

Professor Weizman, in a moving speech, urged the assemblage to stop their flag waving and cheering for Zionism and to dedicate themselves to hard, intensive work for their cause.

"Along with the waving of these flags," said Professor Weizmann, "and the songs of the Hatikvah, another din rings in my ears. I hear the tramp of the Chaluzim (pioneers), and the breaking of stones in the building of roads in Palestine that our great duty may be fulfilled."

More than a dozen times after the meeting started did the audience break forth into the Hatikvah, national anthem of Zion. Judge Julian W. Mack of the Federal district court presided. Others who spoke were Louis Marshall, Judge Irving Lehman of the New York Supreme Court, Rabbi Meyer Berlin, M. M. Ussichkin, Rabbi Abbe Hillel Silver, Shmaiah Levine and Professor Weizman.

Greeting to Sir Herbert

By a rising vote it was decided to send greetings to Sir Herbert Samuel, British lord high commissioner in Palestine. Mention of England as "ever responsive to the cry for justice" brought forth great cheers as did the reading of a message from Justice Louis D. Brandeis of the United States Supreme Court.

"Let us proceed in a little cooler mood than has been prevailing in this meeting," said Professor Weizmann. "What is it which confronts us?

"What is the nature of this trial, and what have we to do in order to pass through this trial with honor and with colors flying? In brief and simple words every Jew is expected to do his utmost toward the upbuilding of Palestine, not only as a refuge of oppressed and poor brethren, but as the home for the Jewish people.

"And because it is going to be the home for those who want it to be their home we shall have to do our very best, and the very best in us is just good enough for the upbuilding of Palestine.

[I will refrain from further quoting Weizmann's speech because there is a more complete transcription in the next article.]

A great outpouring of local Jewry is expected tomorrow night, when Professor Weizman will speak at the 69th Regiment Armory, Lexington avenue and 24th street.

The New York Call, April 11, pp. 1, 2.

5,000 Jews at Metropolitan Opera House Cheer Weizmann and Other Members of Zionist Delegation

Five thousand Jews, waving Zionist and American flags, cheered themselves hoarse, and burst into sharp and flat versions of "Hatikvah" at the Metropolitan Opera House last Sunday when Dr. Chaim Weizmann and the Zionist delegation entered the meeting where New York Jewry had gathered to give them welcome. Every available bit of space had been used, from the platform where those who could not sit were standing, to the side isles [aisles] where hundreds of people who had been waiting in line all morning for admission tickets, stood throughout the entire meeting.

After Judge Julian W. Mack had delivered an address of welcome, he introduced Judge Irving Lehman, Chairman of the American group of the Economic Council, who called upon American Jewry to contribute to the realization of the aspirations of the people who had prayed and suffered for two thousand years.

Dr. Ben Zion Mossensohn, who followed, offered to speak in Hebrew, English or Yiddish a symbol of the versatility of the Jew, said Judge Mack; Rabbi Mayer Berlin, President of the Mizrachi, spoke in Hebrew, as did Mr. Ussischkin.

[There is now a short account of Rabbi Abbe Hillel Silver's address. which I skip because the next article contains a more complete account.]

Dr. Schmarya Levin made one of his typical addresses in Yiddish, full of keen wit and interesting stories. He asked American Jewry to treat this moment seriously and to pay the Jewish debt in deeds, not in words. Louis Marshall followed with a spirited address an extract of which will be found elsewhere on this page.

After another ovation, which lasted several minutes, Dr. Weizmann spoke. In the course of his remarks, he said:

"How is Palestine to be built up? In my humble opinion, there are three lines on which the upbuilding of Palestine is to proceed. The Jewish national home and the Jewish commonwealth which is to follow out of the national home is first of all to stand on the land of Palestine. There is land enough in Palestine at present to contain a population far greater, almost eight to nine times greater, than what it is at present in Palestine. We can for practical purposes bring into the country a vast population without in the least degree encroaching upon the legitimate interests of the present population.

"If you hear to-day about the Arab opposition toward Jewish immigration, it may be a serious and difficult question but it is not a question which is either based on fact or on justice. We Jewish people understand the legitimate aspirations of the Arab nation. But the center of gravity of the Arab national life is not Jerusalem, but Damascus, Cairo, Bagdad. The center of gravity of Jewish life is Jerusalem and is going to remain so.

"We know, and I can give you facts and figures which you will perhaps in a very short time be able to read in the press, pointing out the numbers and quantity of land available. But land is not enough. There must be men and women ready to take over the land and transform it out of the raw material, which it is at present, into what it was thousands of years ago and into what it may become when it is worked and labored upon by modern methods, if we get enough men capable of performing these tasks.

"From my studies and knowledge of Palestine conditions and from whatever we know at present which prevails in the great Jewish communities in the East, we know that to-day there are at least twenty-five, thirty or forty thousand men and women, agricultural workers, well-trained, ready today to stand on the land of Palestine, if given the possibilities. This question whether a Jew can or cannot become an agriculturist has been answered by the work of two generations of Jewish colonization which was first initiated by the father of Jewish colonization, Baron Edmond de Rothschild, and continued by these glorious pioneers who have proved to the world that you can erect Jewish colonies, good Jewish colonies, and so erect them that there should be hundred per cent Jewish labor in it, that from the last stone to the very highest development of this communal life, everything should be done and will be done, and has been done by Jewish work. We know that it can be done, we know that we have the men. What we need at present is the means to combine the men with the land.

Agricultural Colonization Slow

"There is a second important line of development. However successful agricultural colonization may be, it is in its very nature a slow thing. The very success of agricultural colonization depends upon the slowness with which it goes on. You cannot force a plant to grow quicker than it does. You deal here with a factor that may be slightly accelerated by modern methods, but you cannot go beyond certain limits. As I see matters before us, it is of utmost importance that as soon as possible there should be a vast number of Jews in Palestine. I think that we ought to utilize the industrial possibilities of the country and chain up the forces of nature in order to render these forces subservient to the purpose which we have to achieve. In other words, we should utilize what forces there may be in Palestine in order to render these forces a basis for industrial development of the country.

"Schemes are in existence by which the industrial possibilities of the country should be placed before the Jewish public and it will be the duty of the Jewish public to make it possible from tomorrow on to develop what is to be developed in Palestine. I think we shall raise agricultural enterprise to the stage of a science, and not lower the standard of the agriculturist as it is lowered in so many countries.

Chaluzim Want Schools

"The present Chaluzim who come into Palestine ask merely for a tent. The Chaluz asks for elementary tools with which he can work and the next thing he asks for is, have I got enough schools? Or have I enough books to maintain the high intellectual standard to which I am accustomed? For thousands of years this intellectual development was almost the only one which we could carry on unhampered. Within the walls of the Ghetto, when everything outside the walls was raging, our ancestors found strength in the books which they read, and this vast chain of development beginning with the Gaon of Wilna and finishing with Professor Einstein, is nothing but a natural chain, which no community, which no power on earth will break or dares to break. Therefore, along with the material development of the country there must be the intellectual development.

"To-day we come to you and ask for means for colonization. We also ask for means for a Jewish University, which is destined to be the power station which along with the other industrial power stations, is going to keep alive this only power which has kept us alive for thousands of years. I think we have the possibilities, we have the opportunities. We have the human material.

Coming from Every Country

"Our rôle has been merely the rôle of sweepers for those who have got to tread on the road. Who are the people who are going to tread? What is their power of resistance? What is their moral and material power? What are they going to bring into the country? Are they going to be people who are ignorant of the injustices which have been done to us and to them, or will they be people who can survive difficulties and devote themselves as builders of a new old great nation? You probably come from the very districts from which at present these Chaluzim come. You can all imagine the plains of Ukraine and the Jewish youth of good families tramping, tramping, tramping around, beset by every danger which can beset a human being. And they tramp along, nothing on their back; everything which they owned left behind them. All they left is a grave in which is buried everything which was near and dear to them. And they tramp through snow fields under the gray sky of the Ukraine, dangers lurking from every corner. But in their Jewish eyes there is sorrow and tragedy. In their Jewish eyes, Jewish eyes filled with hope, they look towards Roumania as a paradise, because in Roumania there is a port called Constanta and from the port of Constanta there are ships leading to Palestine. And so they tramp, tramp, tramp along. They come from Persia, India, from all the corners of the world. And if they do not come in thousands, it is not their fault. We have not created the proper conditions for them to come in vast numbers.

Ask Only for More Work

"Follow them a little more on the road. They have reached the shores of Palestine and God alone knows no luxuries await them there. Hard and dreary work and the only thing they demand is, 'Is this work secure?' 'Is it going to continue for always?' And these people, upright, straight, educated, courageous, clean looking, fresh air and open air people, they toil on the roads of Palestine and on these roads from Dan to Beersheba, you see and hear to-day Jewish youths singing Jewish songs to the tune of the hammers which break the stones and build the Jewish roads.

"The one question these people ask is, 'Are we going to remain here and look on Palestine, or are we going to be placed on the land? Is there going to be a tie between us and the land?' My answer was, 'I am going out into the Goluth to tell your story, with fire on my tongue if I can do it. I shall sweep the world from one end to the other. I shall start a movement which will strike into the heart of every Jew. I shall make the Jewish streets look as the streets of England looked on August 4th, when no young man or woman who did not wear khaki could show his face. And no Jew who has not contributed his full to the upbuilding of Palestine will be able to show his face. That is the duty to which Louis Marshall referred, who perhaps does not see eye to eye with us. But the Jewish honor, nay, the very existence of Jewry is at stake.'

Take Advantage of Opportunity

"Such an opportunity as you have to-day will perhaps not come back. The price which we had to pay for this opportunity is much too great to desire that it come back. And if you and we miss it, your children will ask you what have you done to miss it. And the nations, and the world will say, 'Indeed, it must be a miserable people who do not utilize an opportunity for which they have prayed, for which they have suffered, and for which thousands and thousands and thousands of their ancestors have suffered.'

Ours Is Key to Palestine

"The key to the doors of Palestine is not in the pocket of Sir Herbert Samuel. It is not in the pocket of the Zionist Organization. It is in your pockets, and it is on you the responsibility rests if we cannot bring in to-day the maximum of Jews which Palestine is capable of absorbing. You American Jews will have to create the conditions. Whereas we ask from you Maaser, perhaps a tenth of your income, the Chaluzim have paid one hundred per cent, and not alone in money. Along with the waving of these flags and the songs of Hatikvah, another din rings in my ears, the tramp of the Chaluzim and the breaking of stones on Palestinian roads. An understanding of this solemn moment is necessary and an appreciation of the fact that we all stand at present at the parting of the ways and eighty generations of suffering, and the great, unbroken tradition of an ancient race looks down upon you American Jewry.

"I believe and I trust that this duty that we are expected to fulfill will be fulfilled honorably and fully and when the fulfillment will come the God of Israel will look upon His people and say; 'I have tried and tested my people. Many of them remained faithful and those who have remained [a line is missing] fathers. And when the march of Jewry to those tents begins, a song of praise, a new psalm will flow from the grand Hermon to the river of Egypt and from the Mediterranean to the Mountains of Moab. The song of the freedom of Israel, the Song of Songs.'"

Although there were repeated calls for Dr. Einstein, the lateness of the hour made an address by him impossible.

Justice Louis D. Brandeis sent an expression of regret to the meeting at his inability to be present. The meeting decided to cable greetings to Palestine to Sir Herbert Samuel, British High Commissioner.

The New Palestine, April 15, pp. 1, 3.

> The maaser is the tithe the Jews gave from the fruits of the soil to the Levites for their service in the temple.

Huge Throng Welcomes Visiting Zionist Leaders

Dr. Weizmann and Others Address Gathering at Metropolitan Opera House, New York—Rabbi A. H. Silver among Speakers—Expresses Belief That Jews of America Will Respond to Palestine Appeal—Now That the World Is Convinced, Jewry Will Be Convinced

NEW YORK, April 11.—(J. C. B. Service)—The Metropolitan Opera House of this city was packed to capacity yesterday afternoon at the official welcome extended by the Zionist Organization of America to Dr. Weizmann, Professor Albert Einstein, M. M. Ussishkin and Dr. Mossinsohn, the four European and Palestinian leaders who are here now on a special delegation to tell American Jewry, the possibilities as well as the needs of the new Zion, the Zion which still has to be built. Judge Julian W. Mack presided and among the speakers were Louis Marshall, Judge Irving Lehman, Rabbi Abba Hillel Silver of Cleveland, Dr. Shmaria Levine and the visitors themselves, with the exception of Professor Einstein, who declined to make a public address.

After explaining the full importance of the occasion, Judge Mack introduced Judge Irving Lehman. He was followed by Rabbi Silver, who in the course of his address said:

"The hour, my friends, does not belong to me nor to any of my colleagues in this country. This hour, I take it, belongs to those men who have come across the sea bearing the message of Palestine redeemed. This throng of tribute-bearing men and women is, to my mind, testimony supreme of the deep sense of gratitude and affection which the Household of Israel living in this blessed land, entertains to that man whose gifts of mind and soul enabled him in God's Providence, to become the Ezra of the second restoration, Chaim Weizmann. We hail and greet his colleagues and coworkers who have come here, that indomitable spirit whose mighty words and whose tremendous message has just now been delivered to you, Menachem Ussishkin. We greet, all of us, that man, that intellectual Titan, who has again given evidence through his labors and his achievements of the intellectual leadership of the sons and daughters of Israel throughout the world, Albert Einstein. And we greet also that man who has given so many years of his life and so much of his splendid gifts and energies to the cause of Hebrew culture and Hebrew education in Palestine, Mossinsohn. I would have them feel, and I am quite sure that I voice the sentiments of everyone here and of the millions of the people who are not here, that we throw open our homes and our hearts in real genuine Jewish hospitality to them. We are blessed for their coming. I would have them feel as they will go through the cities of this land, and when they return to their homes, that we American Jews have come to realize that a new day has come, that the task of tomorrow will differ not alone in magnitude, in size but in quality.

"I would have Mr. Weizmann and his associates when they go back to Palestine, know that our supreme wish at this time shall be the means to continue their labors in Palestine. I would have them bring back from these best throngs of the sons and daughters of our people throughout this land to the Chalutzim, our greetings and our sense of deep gratitude to them, the men who with the sweat of their brow are making rich the soil of Palestine. I would have them feel that we in this land are not unmindful of the supreme sacrifices which they are making. They are removing the reproach from off our people, they are giving answer to the charge that we are not builders, that we are not workers. The Chalutzim are toiling with bleeding hands and torn hands and feet to upbuild Palestine. We cannot match their sacrifice. We cannot do the things they are doing, but we shall try to match them first with our generosity. It has been said that Mashiach Ben David would have come if his hands were not shackled with gold. We are going to break these shackles of gold. We are going to liberate the very spirit of Israel.

"Tell your boys and girls, my friends, tell them that Palestine has need of gifts of their minds and their bodies, that Palestine needs them, for I can but restate in my poorer terms the words so eloquently put by Mr. Ussishkin—we are not building in Palestine a refuge for the outcasts of our people, we are building in Palestine a home for the spirit, for the soul of our people. I am optimistic about the success of the mission of these, our great leaders. Our Jews of America will respond. Now that the world has been convinced, Jewry will be convinced. And I tell you what will help and what will stimulate this mighty conviction in the souls of our people—just that new tide of anti-Semitism that has encroached upon our shores. That is not going to break our spirit and demoralize the ranks of Israel. It will strengthen and crystallize the souls of our people. You know that wonderful saying of our Bible, how wonderfully true it rings at this moment, 'The more you persecute, the more you oppress him, the more he grows and the more he increases.' We are going to remain true to that beautiful symbol of the Midrash, 'Israel is like unto an olive—that fruit of Palestine—.' It does not yield the fruit of its oil until it is crushed, so the Jew does not yield the real oil of his spirit until he is persecuted, until he is oppressed, until the people about him show their resentment and their hatred. We are going to respond to every attack upon our people, to every libel and every slander, by more Jewishness, [by more school, and synagogues, adds the *New Palestine*], by more intensive and loyal work in Palestine."

Mr. Louis Marshall, who was the next speaker, said:

"We must understand one thing. That the time for talking is over. We cannot by means of enthusiasm build up one acre of Palestinian soil. Action is necessary. Liberality and generosity and self-sacrifice are necessary in order to accomplish what you want. One other thing is necessary. You are now facing a world which says the Jews are on trial. The question has arisen as to whether the Jews are able to build up a country, whether they possess a pioneering spirit, whether they are ready to live a life of hardship, a life of sacrifice. One thing is essential to teach the world the lesson that I know we can teach them, if we only chose to do so and that is unity, unity in Israel. Woe to those who bring strife where peace should reign. Woe to those who bring discord where unity should exist. The time has now come when we are to be placed in the scales. I feel that we shall not be wanting."

Dr. Shmaria Levine, who spoke in Yiddish, conveyed to his audience the fact that it was far easier to obtain Palestine for the Jews than to get the Jews for Palestine. Menachem M. Ussishkin, the noted leader of Russian Zionists, as well as Dr. Myer M. Berlin, spoke in Hebrew. The former emphasized the great importance of national work in Palestine and declared that he spoke only in behalf of 80,000 Jews in Palestine, but he felt justified in appealing to the great majority which American Jews form for a partnership in the rebuilding of Zion.

Dr. Weizmann was the last to speak and in the course of his address said:

"You will have to create the conditions and you American Jews, those Jews of yours, many of you the water has not dried up on your coats from crossing the ocean. You know that a double duty rests upon you as Jews and as those Jews who are under the sheltering wing of a great and a generous government capable of living a sheltered life. You must remember that whereas we ask from you Maaser, perhaps a tenth of your income, the Chaluzim, have paid one hundred per cent and not in money but in crying. That is what you will have to remember. And that will be the answer to all those flag wavings to which I am used and which I deeply appreciate. But along with the waving of these flags, and the songs of Hatikvah, another din rings in my ears, the tramp of the Chalutzim and the breaking of stones on Palestinian roads in order to fulfill this great duty. An understanding of this solemn moment is necessary and an appreciation of the fact that we all stand at present at the parting of the ways and eighty generations of suffering and the non-broken grand tradition of an ancient race looks down upon you American Jewry. I believe and I trust that this duty that we are expected to fulfill will be fulfilled honorably and fully and when the fulfillment will come the God of Israel will look upon His people and say: 'I have tried and tested my people. Many of them remained faithful and those who have remained faithful I have brought back into the land of their ancestors.' And when the march of Jewry to those tests begin, then ladies and gentlemen, a song of praise, a new psalm will flow from the grand Hermon to the river of Egypt and from the Mediterranean to the Mountains of Moab. The song of the freedom of Israel, the Song of Songs, and we pray that all of you should here do honor."

The audience was throughout extremely enthusiastic and responsive and the guests received tremendous and lengthy ovations.

The Jewish Independent, April 15, pp. 1, 5.

> "Dr. Weizmann did not qualify as an American orator," Louis Lipsky writes. "His voice was not resonant. He had few gestures. He used no groping introductions or exalted perorations. He hated the impersonation of emotion. He had no ear for the rhythmic phrase. He acquired the English gift for understatement. He did not propagandize himself as a person. He was not made for stage effects.
>
> "In spite of these limitations, no Jewish speaker ever made the same deep and lasting impression—even in the United States. Dr. Weizmann spoke as if his words were the issue of suffering. He made the impression of a murky flame that had to be fanned to give heat. Shmarya Levin had burning passion; . . . Ussishkin took his audience by storm with sledge-hammer blows; . . . Dr. Weizmann had none of these qualities. . . . He seemed to speak ex cathedra for the silent Jewish people. He was their interpreter and advocate. A cause had found a voice for a people emerging from the clouded past and demanding justice from the modern world."[22]
>
> On this very day, Judah L. Magnes, president of the Kehillah of New York City, a pacifist Zionist, proposed that Einstein convene a small meeting of intellectuals interested in the university.[23] For Einstein's answer, see p. 105.

Monday, April 11

Einstein Refuses to Debate Theory

Dean Reuterdahl's Challenge to Discuss Relativity Declined As Detraction from Mission

Dr. Albert Einstein was interviewed yesterday in his headquarters at the Hotel Commodore regarding the attack on his theory made by Dean Arvid Reuterdahl, of St. Thomas College, St. Paul, Minn.

Dr. Einstein smilingly listened to newspaper accounts of the Reuterdahl attack. Through his secretary he said:

"I came here with one object—the promotion of the establishment of the Hebrew University in Jerusalem. I will not be led into a discussion of my theory with persons who may not understand. There may be some personal intent in the remarks of this gentleman, whom I have not the honor of knowing.

"The great purposes of my mission to this country must not be overshadowed by my theory. I will be here a short time, and all of that time must be devoted to the great Palestine reconstruction project.

"I have consented to deliver a few lectures, but beyond that I do not wish to encroach upon my limited time. It must be seen plainly that I cannot enter into newspaper discussions with persons who doubt or misunderstand my theories or question my integrity.

"I have not had opportunity to look into this challenge to debate issued by Dean Reuterdahl. Being without knowledge of the person called 'Kinertia' who is said to have written on the subject, I am not prepared to express any opinion."

It was further said for Dr. Einstein that he had no desire to popularize his theory of relativity; that he had written his book from scientific motives and not for notoriety.

The New York American, April 12, p. 9.

> In his "theory of interdependence," Arvid Reuterdahl, Dean of the Department of Engineering and Architecture at the College of St. Thomas, set up the following syllogism: "Mathematics is an *a priori* science grounded in depths of logical conscious life. Euclidean geometry conforms with the requirements of this logical mental life. It also conforms with the requirements of the external world which is three-dimensional. This is true because there is genuine interaction between the conscious and the unconscious world which are interdependent. . . . [T]herefore the mental world of Euclidean or three-dimensional properties agrees with the physical and external world."[24] Real physical action, such as gravitational force, must explain phenomena and not relativists' warped space.

Tuesday, April 12

Weizmann Pleads for Palestine Aid

Leader of Zionists Says American Jews Must Make It Possible to Reclaim Land—Asks All to Give a Tithe—End of Differences Urged on Supporters of Plan, with Renewed Drive for Funds

A plea for all factions within the Zionist movement to end their differences and get together in the upbuilding of Palestine was made last night by Dr. Chaim Weizmann, President of the World Zionist Organization, before more that 8,000 Jews in the Sixty-ninth Regiment Armory. The occasion was a popular reception to Dr. Weizmann, Professor Albert Einstein, M. M. Ussishkin, Dr. Ben Zion Mossessohn and Dr. Schmarya Levin, members of the Zionist Mission to America, led by Dr. Weizmann.

The armory was filled to its capacity long before 8 o'clock, with the crowd waving the flag of Zion as well as the Stars and Stripes. Outside the armory 3,000 persons waited, hoping for an opportunity to gain admittance.

In presenting Dr. Weizmann, Judge Gustave Hartmann, who presided, gave credit to the leader of the World Zionist Organization for obtaining Palestine as a homeland for the Jewish people under a British mandate. He referred to Professor Einstein as the "master intellect and greatest scientist of the age."

President Harding, in a letter expressing his regret that public engagements made it impossible for him to attend, said of Dr. Weizmann and Professor Einstein:

"Representing as they do leadership in two different realms, their visit must remind people of the great services that the Jewish race have rendered to humanity."

Dr Weizmann said that the hopes of the Jewish people had concentrated on the Jews of America.

"We come to you, my brothers and comrades, to demand and request that you take upon yourselves a great pain of the responsibility and debt which we owe to the Jewish people in Palestine," he said, and added that there were pessimists who did not believe that it would be possible to raise the fund of more than $100,000,000 which the Zionists have set as their goal for the upbuilding of Palestine.

He said that it had been much harder to procure the Balfour declaration permitting this work than it would be to raise the money. In the names of all those who had been waiting through ages for the restoration of their native land, Dr. Weizmann begged of the Americans represented at the gathering to be prepared to give in the old Jewish measure of a tithe.

In pleading for the strengthening of the Zionist movement, Dr. Weizmann said that the moment it was weakened the possibility of developing Palestine would become endangered.

"The pioneers cannot wait," he continued. "They are already on their way. We must accomplish this mission for them, because if we do not, they will come to us with a very grievous complaint, and they will say: 'Why did you bicker? Why did you have bickerings at the time when we, thousands of us, tramped through dangers of all sorts and through hazards of all kinds?' "

Professor Einstein made the briefest speech of the evening, when he said:

"Your leader, Dr. Weizman, has spoken, and he has spoken very well for us all. Follow him and you will do well. That is all I have to say."

President Nicholas Murray Rutler of Columbia, in welcoming the visitors, said that it was remarkable to observe that during the greatest war the world has ever seen "these men of science went on dealing with the deepest and profoundest thought as if there were no noise going on in the world."

The New York Times, April 13, p. 13.

> The meeting did not go as smoothly as it looks, because Ussishkin was also among the orators, and "he acted wretchedly by his intolerable speech," complained Stephen S. Wise to Weizmann when they had lunch on April 26.[25]
>
> Ussishkin, "this man of granite seemed made to rule," Louis Lipsky characterizes him. "That was what he thought, too. He had the nature of a czar whose opinions issued in the form of edicts. He was deadsure that he was always right and no one else could be as right as he."[26]

8

Demonstrates with Chalk

✦

Friday, April 15

Einstein in Lecture Explains His Theory

Professor Demonstrates with Chalk As Audience in Horace Mann School Applauds—It's a Theory of Method—Experience Supplies the Postulates and Reasoning Draws the Conclusions, He Asserts

Professor Albert Einstein lectured on his theory of relativity yesterday for the first time since his arrival in this country before members of the science faculties of Columbia University, and students, in the auditorium of Horace Mann School. He spoke in German, but those anxious to see and hear the man who has contributed a new theory of space and time and motion to scientific conceptions of the universe, filled every seat and stood in the aisles.

Professor Einstein was greeted with great applause when he entered the room escorted by Professor M. I. Pupin, Professor George B. Peagram, and Dean Woodbridge of the science schools. Professor Pupin introduced him as the discoverer of a theory which is "an evolution, not a revolution of the science of dynamics."

In discussing the abstract conceptions of his theory, Professor Einstein showed himself possessed of a sense of humor, for he several times brought chuckles and laughs from his audience by his references to the "idiotic" behavior of certain bodies in accelerated systems. Also he caused much amusement when he wished to erase some diagrams he had drawn on the blackboard and made futile motions in the air with his hand until Professor Pupin came to his rescue. But for the most part, his audience listened to him with the absorption of men of science listening to a brother scientist expound a theory which may alter all their conceptions of motion and space.

Not a Closed Theory

"The relativity theory is not a closed theory, but a theory of method in which experience supplies the postulates and reasoning draws the conclusions," said Professor Einstein. Special relativity, he said, endeavored to find laws to account for the inability to detect absolute motion. From a purely descriptive point of view every motion was relative. It was impossible to describe the motion of a point except by its relation to another point. If two systems were moving past each other uniformly, either might be looked upon by the observer on one as at rest, with the other in motion.

The basis of relativity was Newton's first law of motion that a body moved in a straight line with uniform velocity when far away from other bodies. Professor Einstein set up two co-ordinate systems, which he sketched on the blackboard to show that the motion which one particle in a system bears to another particle in another system may result in a curve. In ordinary mechanics only one system was permissible, the second system being permissible only when it moved with uniform motion relative to the first.

The mechanics of Newton and Galileo embodied the special relativity principle, he said. There was no cause for doubt of this law in the mechanical field until electric and optical phenomena were examined.

Refers to Maxwell's Theory

Light traveled in a straight line with uniform velocity in space, said Professor Einstein, referring to Maxwell's theory of optical phenomena. He inquired in what co-ordinate system Maxwell's laws held, and how might the system be moving? If one co-ordinate system, which he sketched, moved with respect to the other, it should be expected that the velocity would be less if the second system moved in the direction of light. If this were true special relativity would be violated because absolute motion could be detected.

Michelson's experiment, however, showed that the velocity did not change and the special relativity principle held in electrical phenomena as well as in pure mechanics. This introduced a theoretical difficulty into electromagnetic theory, as formulated by Maxwell, who predicted change.

Relativity doubted "simultaneity" where applied to two events in different places. There must be a method of determining simultaneity. A physical relation between the events was necessary. Relativity relinquished the absoluteness of simultaneity. In explaining this, Professor Einstein drew a line to represent a rod, its middle point to be determined by measurement. Simultaneous events took place at the ends when seen together from the middle, thus furnishing a criterion of simultaneity. To find the centre there must be a measure. Simultaneity held only for the special co-ordinate system; if it was simultaneous for one system, it was not for the other, because by the time the light reached the observer at the middle of the rod he would have moved with respect to the other system.

Relations Between Space and Time

In stating the mathematical relations between space and time co-ordinates of two systems, so as to reconcile the special relativity and the constancy of the velocity of light, in order to formulate physical laws, Professor Einstein again reverted to his example of the rod. A shortening of a rod in the direction of the motion, as it could be measured by a stationary measure, would do this, he said. This appeared contradictory until the change in time was taken into account.

Clocks on the moving rod go slower. To prove this he drew a rod with mirrors at each end, the rod moving in a direction at right angles to its length. As the rod moved and light was reflected from the mirrors, the light moved in a diagonal course between the mirrors, caused by the forward motion. This diagonal course was longer than the direct course, therefore the clock, pictured by the mirrors, went slower. All clocks on bodies in motion went slower, he said. Another consequence was that the energy associated with a body determined its inertia, and led him to lay down the law that "inertia is nothing but energy." The method and consequences of the special relativity theory involves a uniform relative motion.

A special relativity formulation did not hold for a co-ordinate system moving arbitrarily. But such a formulation must be possible. The difficulty with Galileo's theory was in the first law stated that a body moved in a straight line when no others were present, and no others were present when it moved in a straight line, an argument in a circle.

When Galileo's Law Fails

Professor Einstein inquired as to the mechanics in an accelerated system. Galileo's law, he showed, did not hold there. In a stationary co-ordinate system a body would keep zero velocity if it had no motion at the start, but in the second system it would have acceleration. Just what happened near material bodies, such as the earth? He spoke of the phenomena of falling bodies, and showed that they behaved just as if we were in an accelerated system.

He showed the behavior of rods in such a system. With a number of equal rods we could build geometrical forms, and have always assumed that such construction obeyed Euclidian geometry. But, he pointed out, to the amusement of his audience, in an accelerated system they behaved "crazily." In an accelerated system there was a gravitational field, and the rods in a gravitational field behaved as in an accelerated field. We can now set up a gravitational theory based on this equivalence which yielded Newton's law of gravitation.

The New York Times, April 16, p. 11.

> Even though this account is inaccurate, sometimes incorrect, it is amusing. Those familiar with special relativity will enjoy how reporters are stumbling on a terrain too rugged for them; for interested readers I recommend Einstein's popular exposition, *Relativity: The Special and the General Theory.*

Einstein's Reception

By Edwin E. Slosson

To the student of mass psychology, such as all of us are, there is nothing more interesting than to observe the popular reaction toward a new idea. Emerson says: "Beware when God lets loose a thinker in this planet," or something of the sort. I am not sure of the wording for I am writing in a railway station and, not being a Bostonian, I have not taken Emerson along in my suitcase.

Einstein is undeniably a thinker. Whether he is a right thinker or a wrong thinker remains to be proved. That does not matter to the social psychologist who is only interested in the effect produced by dropping a new thought into an ocean of old ones. It is like dropping a crystal into a solution of some salt. Sometimes the crystal is dissolved in the liquid and disappears. Sometimes it sets the whole mass to crystallizing in its own image. Sometimes it lies on the bottom inert, neither dissolving itself nor solidifying the solution. Whether the crystal is absorbed or rejected or produces an immediate reaction depends upon when it is dropped in. Likewise the effect of a new idea on the world of thought depends upon its coming at the psychological moment.

Einstein's idea was a war-baby. While the cannon were booming, his brain was laboring and during the noisy years of 1915–1917 his generalized theory of relativity was brought forth.

But at that time the wires were down between this country and Germany. Most Americans thought it did not matter much because they read repeatedly in the papers that German science was a fraud and that all the really important discoveries had originated in other countries, altho it was admitted that the Germans had an uncanny knack of stealing these ideas and putting them to profit. It was almost impossible for an American scientist or an American editor to get a German publication. In one of the states a bill was introduced into the legislature requiring the universities to burn all their German books.

The British did not adopt this ostrich-like policy. They kept their eyes open on all that was doing in enemy countries, whether scientific, technological or political. British editors were kept supplied with German periodicals and for the benefit of those who could not read German the British War Office published very complete and impartial abstracts of the enemy press. While the war was at its height British astronomers were studying Einstein's work and preparing expeditions to be sent to Africa and South America to test his predictions at the next eclipse. Professor Eddington of Cambridge published a thoro report on the relativity theory of gravitation in 1918, and the photographs he took of the eclipse of May 29, 1919, confirmed the calculations of Einstein.

But at that time many American scientists were still in the dark as to what had been done in Germany during the war and had not yet got access to Einstein's later papers. So when the reporters came around to interview them on what Einstein's theory of relativity was and how the eclipse observations confirmed it, some of them were puzzled to reply since they knew only of the special theory of relativity which Einstein had put forth in 1905 and which had nothing to do with gravitation.

The French like the British kept up with the progress of German science during the war, altho unlike the British they took no stock in Einstein. Alphonse Berget in *Larousse mensuel* of 1917 makes fun of the Germans for taking to relativity. He says that they are fond of drawing the long bow and have "forgotten that common sense is the necessary basis of science."

True science is above all barriers of nation, race or sect. Einstein himself, altho he is an ardent nationalist and comes to America as an advocate of Zionism, is nevertheless a sincere internationalist and welcomes all efforts to reëstablish the world commonwealth of science. He expresses this hope in his letter to President Butler for the Barnard medal which Columbia University has awarded to him and in his letter to the London *Times* of December 13, 1919, he makes a witty reference to the present tendency to discriminate between men of science on the ground of nationality:

"The description of me and my circumstances in *The Times* shows an amusing feat of imagination on the part of the writer. By an application of the theory of relativity to the taste of readers, today in Germany I am called a German man of science and in England I am represented as a Swiss Jew. If I come to be regarded as a *bête noire,* the description will be reversed and I shall become a Swiss Jew for the Germans and a German man of science for the English."

And so indeed it has turned out since.

It is unfortunate that Einstein should make his first appearance in America as a Zionist instead of a scientist. He would have done more for Judaism in general and for Zionism in particular if he had come to America, like Bergson, as one of the great thinkers of the modern world whom all America delighted to honor rather than as a leader in a separatist movement of a race. But so long as there are some who hold the Jews as a whole responsible for those of their race who have itching palms or dirty fingernails, we must expect them to show in return a disposition to monopolize their men of genius. Since the Jews have more men of genius in proportion to their numbers than any other race since the ancient Greeks, they could afford to be generous with them. It is to be hoped that Einstein's theories will be judged objectively on their merits, without prejudice in their favor or against on the ground of race.

The United States owes peculiar gratitude to Professor Einstein, for he has answered two American conundrums. Simon Newcomb, the American astronomer, pointed out in 1895 that Newton's law of gravitation failed to account for the shifting of the orbit of Neptune. It was Abraham Michelson, the American physicist, who in 1887 proved that no evidence of ether-drift thru matter could be detected. Einstein's theory accounts for both these puzzles besides giving the clue to several others.

The application of Einstein's theory of relativity to astronomy and physics is as revolutionary as the theories of Copernicus and Newton. What will be the result of its application to chemistry and biology we cannot yet imagine, but evidently the effect will be still more fundamental. For Einstein's laws of motion do not differ appreciably from those of Newton except at extremely high speeds approaching the speed of light. This is far beyond the possible speed limit of any body that the astronomer or physicist has to deal with. But the unit of the chemist, the atom, is now believed to consist of a nucleus of positive electricity around which revolve one or more negative electrons with a speed almost that of light. Sometimes they fly off on a tangent as in the case of radium. These same swift electrons are also supposed to play a part in the processes of life. For instance, one theory of the growth of plants is that the shorter light-waves from the sun shake loose the electrons from the chlorophyll in the green leaf and this provides the energy for the building up of protoplasm. So both chemical and biological reactions may come within the realm where Einstein's laws rule instead of Newton's. And Einstein's laws of motion are extremely upsetting to our ordinary notions! If chemists and biologists have to take up four-dimensional mathematics we may expect an exodus from those sciences into sociology and psychology where as yet nothing more than arithmetic is required.

In their reaction toward a novel idea people may be divided into two classes, the neophiles and the neophobes, those who are drawn to it because it is new and those who are repelled by it for the same reason. Then there are always some who say it is true but not new, and others who say it is new but not true. Of course the humorists, who always stand as sharp-shooters on the outer ramparts of conservatism, direct the arrows of their wit against any stranger seen approaching. Einstein is received with the same chorus of laughter as greeted Galileo when he said the earth moved; Darwin when he talked of evolution; Columbus when he proposed to get east by going west; Stephenson when he asserted that his locomotive might catch up with a cow; and the hero who first dared carry an umbrella in the London street. Wit is a defensive mechanism for society as well as for the individual and serves a good purpose on the whole since for one worthy man who has suffered like these a hundred cranks and pretenders have been laughed out of court.

But it does not matter, one way or the other, what the popular reception of Einstein's ideas may be, for they will be weighed in the balance by the few who are competent to pass upon their value. They are now the chief topic of discussion in the field of physics and new ways of testing them are being devised.

The Independent, April 16, pp. 400–401.

> It was not Neptune but Mercury whose irregular motion was a problem to solve; it was Urban J. J. LeVerrier who in 1859 discovered the anomalous motion of Mercury, Simon Newcomb only added more to LeVerrier's data on the motion of planets; the quote from Einstein is not from his letter of December 13 to the editor of the *Times,* because there was no such letter, but from Einstein's article published there on November 28.[27] Please ignore how the (shell) electrons' leaving the atom is conflated with radioactivity; but even this is pretty good for a "student of mass psychology" and a "social psychologist" who has as yet been required to learn "nothing more than arithmetic."

Aladdin Einstein

The popular interest in America in Professor Einstein's theories has astonished the professor. The public who does not know whether the theory of relativity has accounted for the alteration of mercury or of Mercury, waylays his steps, and delights, with the exception of a mere alderman or two, to do him honour. Gifted newspaper-reporters herald him as the originator of the theory of relativity, which, by the way he is not, and question him as to the ultimate nature of space, though only a mathematical physicist who is also a philosopher could understand the professor's answers.

This general interest in an extremely difficult science is not quite what it seems. Probably Professor Einstein does not realize how sensationally and cunningly he has been advertised. From the point of view of awakening popular curiosity, his press-notices could hardly have been improved. The newspapers first announced his discovery as revolutionizing science. This sounds well, but its meaning, after all, is rather vague. Then they printed a series of entertaining oddities, supposedly deducible from his hypothesis, although most of them could have been equally well deduced from the conclusions of Lorentz or Poincaré: for example, moving objects are shortened in the direction of their motion. This is a gay novelty until one learns the proportion of the reduction, which is calculated to divest the statement of interest to any but scientists. Further, our newspapers told us that if we were to travel from the earth with the speed of light, and could read the clock we left behind, it would always remain at the same moment, permanently pausing, unable to reach the next tick. But we should be unable to travel at the rate of light for a number of reasons, the most interesting and perhaps the most decisive being that such a speed would cause our mass to be infinite! Finally, our informants assert that no point in space, no moment of time can serve as a permanent base for measurement; we can measure only the relations of space, the relations of time, never absolute space or time; and even to measure space-relations, we have to take into account time! What a fascinating dervish-dance of what we used to regard as immutable fixities! Is it possible that these delicious contradictions are serious and accredited doctrines among those who know? Yet so they appear, for though Professor Einstein is always careful in stating that his hypothesis enjoys as yet only a tentative security, his methods are vouched for by the experts, his procedure is according to Hoyle, and the crowd is at liberty to gorge its appetite for marvels untroubled by the ogres of scientific orthodoxy.

Aside from the fact that Professor Einstein comes as a distinguished and somewhat mysterious foreigner to partake of our insatiable hospitality, his popular welcome is to be accounted for by the spell of wizardry that the press has cast upon his interpretations. For it is the necromancy of these strange theories, not their science, that catches the gaping crowd. Reporters are often good, practical psychologists. Instinctively they have divined the public eagerness for miracles, without grasping the factors that feed this taste. They know that most of us are essentially children still clamouring for fairy tales. Man is congenitally restless with the prison-house of this too, too solid world. He is always looking for short-cuts to power. Since he can not find them to his mental satisfaction as once he could through the miracles and divine dispensations of the Church, or through the magic and occultism that were his legitimate resources in the Middle Ages, he now turns to the wonders of science and philosophy. Here, even in theories that he does not understand, he can find release for his cramped position, here he can taste the intoxicating freedom of a boundless universe, and renew his sense of personal potency. It is therefore with a curious elation that we greet this bold explorer of the swiftest speeds, the vastest spaces, the remotest times that our universe has yet yielded. We bestow upon him the keys of our cities because he has accorded us the freedom of the cosmos. He has quickened our imaginations so that we leap upon a ray of light, escaping time. Or starting from some distant star

> Whose light, through the unplumbed abyss
> Falls, ere it reach us, through a thousand years,

we sweep earthward by the straightest path that anything physical may keep, only to reach our planetary system and bend ever so slightly towards the margins of the sun! We find in relativity a strange and literal confirmation of the egoistic assertion that man is the measure of all things. If we are at all familiar with the theories of the nature of matter, we gain through our visitor's conception of mass and inertia, a more comprehensive simplicity, running through all forms of energy from stocks and stones to electricity and light. We are awed by the beauty and terror of this more pervasive order, this deeper music of the spheres.

But, after all, is the impulse that produces the scientists themselves so different? Something more of curiosity they must have, than the common run of us, something more of the patience permitting an exquisite conformity to the laws of their subject-matter, more, perhaps of that creative intelligence which allows at once a greater scope and a more concentrated intensiveness of inquiry, but is not their devotion to the truth also determined by its power to make them free? For "pure" scientists as well as laymen, it is quite as much the poetry as the utility of science that lures, and the impulse of the race searching an outlet into a vaster, more harmonious, less finished existence, answers with an Einstein.
Gertrude Besse King

The Freeman, April 27, pp. 153–54.

Monday, April 18

Rabbi Wise, Pledges Amount for Einstein's Palestine Project

Stephen S. Wise, pledged a contribution of $10,000 to Professor Albert Einstein for the building of a Jewish university in Palestine, in the Hotel Pennsylvania, last evening, when the synagogue celebrated its fourteenth anniversary. Professor Einstein said the proposed university would be "the greatest thing in Palestine since the destruction of the Temple of Jerusalem.

"I became interested in the undertaking," he said, "after many students had been refused admittance to the University of Berlin because of their Jewish nationality. One of my colleagues and I started classes for these young men, but their numbers soon grew far too large for us to handle. Thus the idea of a special university for Jews appealed to me as meeting a special need."

. . . .

The New York Times, April 19, p. 6.

> Einstein is here talking about the "courses of Berlin university professors accredited by the state," shortened by the press to "university courses for Jews," that he and Leopold Landau organized for young Eastern Jews in Berlin in 1920.[28]

9

Defines the Speed of Light

✦

Monday, April 18

Einstein Defines the Speed of Light

Says Its Velocity Is Relatively Variable Based Upon Whether System Is in Motion—He Also Discusses Time—The Faster the System Moves the More Slowly Time Passes, He Tells N.Y. College Faculty

Professor Albert Einstein delivered the first of a series of four lectures at the College of the City of New York yesterday, before the Faculty and their guests. He paralleled to a large extent his lecture of last week at Columbia University, although in developing his theory of special relativity he went into much more detail than before and stated that a rod moving with the velocity of light would shrink till it disappeared.

Mathematics was the principal method by which he explained his theory, and he covered the blackboard with formulae. Apparently those who heard him have a preference for higher mathematics, for when he asked them at the end of his lecture if they wished him to develop his theory mathematically or proceed to a more general discussion of relativity they voted for the formulae.

President Sydney E. Mezes introduced Professor Einstein and expressed the appreciation of the college Faculty for the opportunity of hearing him. He referred to the relativity theory as "reshaping our concept of the universe in its furthest reaches and most elusive parts." Professor Einstein apologized for speaking in German.

Velocity of Light Variable

The special relativity theory arose from the question of whether light had an invariable velocity in free space, he said. The velocity of light could only be measured relative to a body or a co-ordinate system. He sketched a co-ordinate system K to which light had the velocity C. Whether the system was in motion or not was the fundamental principle. This had been developed through the researches of Maxwell and Lorentz, the principle of the constancy of the velocity of light having been based on many of their experiments. But did it hold for only one system? he asked.

He gave the example of a street and a vehicle moving on that street. If the velocity of light was C for the street was it also C for the vehicle? If a second co-ordinate system K was introduced, moving with the velocity V, did light have the velocity of C here? When the light traveled the system moved with it, so it would appear that light moved slower and the principle apparently did not hold.

Many famous experiments had been made on this point. Michelson showed that relative to the moving co-ordinate system K1, the light traveled with the same velocity as relative to K, which is contrary to the above observation. How could this be reconciled? Professor Einstein asked.

How to Define Relative Time

The answer lay, he said, in the dependency of time on the co-ordinate system. To define time with relation to the co-ordinate system, a clock was put at the origin of the system, it being possible to fix the time of any event at its origin. An observer in the vicinity of the clock must, in judging the time of the event, at his position, take account of the time for light to reach him. A clock must be at the event, and the clocks must be synchronized by using the constancy of the velocity of light to obtain a definition of simultaneity. If an observer midway between the two events saw them simultaneously, they were simultaneous. This served as a basis for a definition of time relative to co-ordinate systems. The clocks might be regulated in this way, but the clocks must be fixed in a co-ordinate system and this determination served only to synchronize the clocks in only one system, K. But when two events were simultaneous in K they were not simultaneous in K1.

Professor Einstein then used his example of the rod to determine that the concept of simultaneity was not absolute. Next he considered length, which was easy when the rod was at rest. But when it was in motion it was more difficult. It was easy for a person on a moving rod to find the length, but it might be necessary to measure it with apparatus fixed on K. He determined the beginning and the end of the rod with reference to K at the same time, and measured the distance between them on K. Time entered through simultaneity and, a priori, it was not possible to know whether the length was the same measured from the rod and measured from K. It could be investigated logically, however. The independence of time and length, and motion, had always been assumed, but these ideas must now be dropped, he said.

Where Newtonian Law Fails

Newtonian mechanical law held only for certain co-ordinate systems which were called inertial systems. It did not hold for accelerated systems. If K was an inertial system, then a system K1 was an inertial system if it had uniform motion with respect to K. This was the special relativity principle. The same principle held for light.

Professor Einstein then demonstrated what conclusions these transforming equations led to with regard to space, remembering that there was no absolute measure of time and space. The physical measurements in the same system were not affected by the motion of that system.

Professor Einstein used mathematics to demonstrate the relations between moving bodies. He said that time passed more slowly the faster the system moved. The constancy of the velocity of light forced us to these conclusions by mathematical processes which admitted of no doubt, although direct experimental verification was not possible.

The New York Times, April 19, p. 17.

> This was also the day when Einstein answered Judah L. Magnes's letter of April 10. He proposed a gathering not exclusively of intellectuals but also non-intellectual but "influential" personalities.[29] Naturally understanding this euphemism to stand for rich people, Magnes withdrew his proposal on April 19, saying that he did not have fundraising in mind.[30]

Wednesday, April 20

Calls Density Key to Size of Universe

Prof. Einstein Says If It Does Not Approach Zero Then the Universe is Finite—But Can't Prove Theory—Astrophysical Experiments Unable as Yet to Determine Necessary Data, Scientist Asserts

Professor Albert Einstein touched on the question of a finite universe in his lecture yesterday at the College of the City of New York. He did not say that the universe was finite, for astrophysical experiments had not yet been able to determine the data by which such a statement might be proved, but, as explained some months ago in *The New York Times,* he said, that if it were found that the average density of matter in the universe did not approach zero, then the conclusion could be drawn that the universe was finite.

His lecture was heard by the largest group of scientists and students who have yet attended his lectures. Among the former were several members of the Columbia physics and mathematics Faculty, including Professor Kasner, Professor G. B. Pegram, Professor A. P. Wills, Professor G. V. Wendell, Professor H. W. Webb and President Mezes of the City College, with many members of his staff. They listened to Professor Einstein's exposition of general relativity as distinguished from special relativity to which he had hitherto confined his remarks, and were particularly interested in his passing reference to the limitation of the universe, which came at the end of his talk.

Euclidean Space Infinite

Euclidean space was infinite, he said, but an infinite universe was only possible if the average density of matter in the universe approached zero as it was extended. If there were stars everywhere, if the density of matter did not decrease so that the average density did not approach zero, then the conclusion must be that the world was finite.

In special relativity it was found that there was no unique co-ordinate system with reference to which natural laws had to be described, Professor Einstein said, but that it was merely necessary that the co-ordinate system be an inertial system. Galilean law was that in an inertial system a material point far from other matter moved uniformly in a straight line. But viewed from an accelerated system the point would describe a parabole. So that

Galilean law did not hold for an accelerated system, hence it must be abandoned in developing the theory of generalized relativity.

There was, however, another general law, that of the equal acceleration of all bodies in the gravitational field which led to the equality of the gravitational mass of a body (as measured by gravitational attraction) and the inertial mass (as measured by the acceleration caused in it by a given force). The agreement of these two quantities was most important; and the weakness of classical mechanics lay in its failure to give the proper significance to this fact.

Modifies Conception of Mechanics

Reverting to the accelerated co-ordinate system, it was seen that all bodies in it had the same acceleration, and the behavior of free bodies in such a system was identical to their behavior in a gravitational field.

"Now," said Professor Einstein, "we are in a position to modify our conception of mechanics so as to apply to any co-ordinate system, for we can investigate an accelerated system by simply using a suitable gravitational field in connection with it and forgetting the acceleration."

The inertial mass of a body which was in an inertial system manifested itself as the gravitational weight of the body when observed from an accelerated system. This equivalence of gravitational and inertial mass led Einstein to consider the possibility of developing a generalized relativity theory. By changing the conditions of observation it was possible to create a gravitational field, and by investigating the gravitational field created being led to gravitational law. The task of generalized relativity was to replace a gravitational field by the suitable co-ordinate system, find the natural laws, and compare them with actual experimental law.

Professor Einstein took the blackboard as a co-ordinate system and asked his audience to imagine a second co-ordinate system rotating about an axis perpendicular to the blackboard.

Euclidean Geometry Not Valid

"Does Euclidean geometry and measurement of time as in special relativity hold here?" he asked. "Look at the situation either from the blackboard or the rotating system. Take measuring rods which are equal when at rest and lay them along the radius of the disk in the rotating system from the centre to the circumference, and around the circumference. According to special relativity, the rods on the circumference become shorter while the rods along the diameter, having no motion in the direction of their length, do not shorten. When the disk is stationary the relative number of rods on the circumference and diameter is pi, about 3.14, the ratio given by Euclidean geometry. But when the disk rotates more rods will fit around the circumference, and the relative number is no longer pi. Hence Euclidean geometry is not valid here. In view of the equivalence of the gravitational field and the accelerated system, Euclidean geometry must be rejected for the gravitational field also.

"Likewise with time. A clock on the periphery moves slower than a clock at the centre for an observer on the blackboard. It must go more slowly, as has been shown, for the observer on the disk, hence the conclusion that as soon as a gravitational field is at hand, of which this is a special case, then clocks at different places run at different speeds.

How to Deal with Space and Time

"On such a disk and therefore in a gravitational field, we cannot build up a cubical system with rods, for this is Euclidean. This means that we cannot use Cartesian co-ordinates, which depend on Euclidean geometry. How, then, can space and time be dealt with? Mathematics has long known the necessary theory, the geometry of a curved surface investigated by Gauss, that is, the behavior of little sticks arranged on such a surface yields the method when the geometry is generalized to four dimensions. While Euclidean geometry fails on such a surface, it can be applied locally to a sufficiently small portion of the surface."

As elaborated by Riemann and Neumann, this geometry is used by Einstein to describe physical space and time.

The New York Times, April 21, p. 12.

Aside from Riemann, I know only one "mann" who helped Einstein learn non-Euclidean geometry, but it was Grossmann, not Neumann.

Thursday, April 21

Einstein Explains Fallacy of Ether

Space without Electromagnetic Fields Has No Characteristics, He Tells Students—Humorous at City College—Talks about "the Greatest Pain of Physics" in Exposition of His Relativity Theory

Dr. Albert Einstein showed yesterday in his lecture in the College of the City of New York how his theory of special relativity does away with the ether in absolute space. The quanta theory, or the theory that energy is given out in parcels and is not continuous, is illogical, he said, whereas the relativity theory is quite logical in its development.

"This is the greatest pain of physics today," said Professor Einstein humorously. During the course of his lecture he frequently smiled as he pointed out some of the amusing results of the acceptance of the relativity theory.

Professor Einstein defined light-ether, which is the medium for the transfer of light in the mechanical theory of light radiation. If it is to transfer energy in this way it must have the material properties of inertia and elasticity, he said. The electromagnetic theory pushed this view into the background, and it came to be considered that only such a field was necessary. This difference in the point of view, he said, was brought about particularly by Lorentz's theory which followed the work of Hertz, and finally all mechanical characteristics of the ether were lost.

In mechanics there was something similar to the conception of ether, he pointed out. Newton was responsible for the abstract character of space as understood in classical mechanics. Besides the inertia system which may serve for reference in classical mechanics, something more basic is required to refer things to, something known as the ether or absolute space, he continued.

"Special Relativity's" Action

The effect of the special relativity theory in destroying the idea of absolute rest and showing the equivalence of all inertia systems is to cast off the ether in the sense of absolute space. That is what is meant by "special relativity doing away with the ether," he said. If physical properties are ascribed to empty space we speak of an ether. But space without electromagnetic fields has no characteristics whatsoever. On this is based the special relativity theory that the electromagnetic field is the ultimate reality, and it is only where electromagnetic fields are found that kinematic or material properties exist.

The quantum theory grew out of the radiation theory, he said. A theory can be developed on classical lines, but the theory suffers from self-inconsistency and it does not correspond to actual experiment, was his contention. This theory leads to the conclusion that the electromagnetic energy of the universe, the energy of light waves, wireless waves and X-rays, accumulates among the higher frequencies, that is, among X-rays and still shorter rays. But this is not so, Professor Einstein said. Max Planck formulated a theory which does find verification in fact, and he gave a physical meaning to his expressions.

A new physical constant had to be introduced with peculiar origin, he went on. Planck investigated the number of ways that energy can be distributed between different frequencies, to do which a certain discreteness must be attributed to energy, said Professor Einstein. Planck assumed that the energy in radiation occurred in amounts which were proportional to the frequency of the radiation. This is a fundamental principle in molecular constitution.

Classical theory agrees poorly with experience, he said. For instance, sunlight causes certain chemical decompositions. Looked at from the viewpoint of classical mechanics, the radiation of electromagnetic fields shakes the molecules apart by violence. So if light is weak enough, no decomposition should be expected. But, however weak the light may be, he said, as long as it is of the right frequency the decomposition takes place. No matter how intense the light may be, the decomposition does not happen if the light is not of the right frequency.

How Planck's Theory Works

A continuous distribution of energy among all wave lengths is necessary to explain interference phenomena, a conflict which constitutes the pain of physics, Professor Einstein had previously referred to. In Planck's theory a material system can only assume certain energy values, that is, a material system can only assume certain states, a, b, c, &c. The relative probability of these states is what must be investigated, he said.

That the quality of light from a material system, that is, the frequency radiated, depends absolutely upon the energy in the emission is a fundamental assumption, he added. The frequency of emission is unrelated to the frequency of vibration of the material system before and after emission.

Professor Einstein examined a material system in which only two states, A and B, are possible. The system can go from B, which is the state of greater energy, to A by radiating energy. If such a system be put into a field of radiation this field through its action may cause it either to go from B to A or from A to B. The probability that the system will go from A to B under this influence is found, and may be compared with the probability that the system will go from B to A, due to simple radiation and to the effect of the field.

But consideration also must be given to the equilibrium between the number of systems having state B and those entering state B. Thus Planck's law of distribution of energy is derived. If it is true that light is absorbed and emitted in such quanta as Planck finds Professor Einstein said, it must be considered that the emission of light takes place in one direction from a resonator instead of spreading out in all directions like a sphere.

The New York Times, April 22, p. 12.

Prof. Einstein Given Ovation

Discoverer of Relativity Theory Makes Short Speech in German at Chapel— Thanks College for Welcome—In College Picture—Climax of the Professor's Visits to C.C.N.Y.—In Special Interview with Reporter Praises Co-operative Spirit of Students and Professors

In an ovation such as the College seldom witnesses the world-famous physicist, Zionist and philosopher, Doctor Albert Einstein, was welcomed last Thursday at Chapel. After many days of unflagging effort on the part of the College authorities, President Mezes, Professor Goldsmith and Dr. Wetzel were finally successful in persuading Professor Einstein to address the student body.

In his introduction Dr. Mezes cited the distinguished and honorable guest as a man "who is considered as among the greatest of the five or six illustrious scientists of the world." A burst of applause greeted the words of the President, and continued even more violently when Dr. Einstein himself rose from his chair and walked slowly across the platform. After a few more seconds the acclamation weakened and finally died out. The voice of the professor re-echoed clearly throughout the Great Hall as he made his laconic address in German.

"It gives me the greatest pleasure, on this, my first visit to America, to have this opportunity of meeting the student body of this great university. I appreciate very much your friendly reception and applause and extend to you all my heartiest good wishes."

A "Big Varsity," led by Ken Nunes, with three big Einsteins on the end, followed, and then Dean Brownson informed the assembly that a picture of all the students and faculty would be taken immediately after the adjourning of Chapel.

President Mezes and Dr. Brownson led Professor Einstein to the Stadium, and Professor Goldsmith and Dr. Wetzel followed with Mrs. Einstein. The crowd of curious students surged around the dignified procession and attempted to draw still closer, so as to get a good view of the scientist.

While on the field, Professor Baskerville offered Dr. Einstein a cigar, and Mrs. Einstein scrutinized the cigar closely and then asked Professor Baskerville, "Ist es veilleicht zu stark fur mein Mann" (It is perhaps too strong for my husband).

At first sight Dr. Einstein seems to be a great musician, rather than a great scientist. His long black hair, combed in the fashion of a Lizzt, his high forehead, his dreamy eyes and dark complexion all belie the man of learning, the deep and penetrating thinker, the man of boundless power and resource.

The address at Chapel was the fifth of Professor Einstein's in the College of the City of New York. He had spoken to a select audience of professors two weeks ago. Last week he delivered four lectures, one each afternoon, on the first four days of the week. Some of the greatest and most prominent scientists in New York were present at these lectures. Only a handful of students was admitted, although a curious crowd attempted to force its way in. He spoke in Doremus Hall, which was amply filled. The subjects of his addresses was his own theory, so little understood by the world. Of all those present, only Professor Cohen could summon enough courage to ask the speaker a question regarding the theory of relativity. A great many of the audience were confounded and mystified by the deep mathematics and strange inductions of Professor Einstein. On one occasion the professor was literally pushed into Professor Baskerville's private staircase to prevent his being crushed by the surging multitude. City College is the only institu-

Einstein in the field of Lewinsohn Stadium of City College of New York, with Elsa Einstein, President Mezes, faculty members, and students.
Courtesy Thompson Photo Company, Rahway, New Jersey.

tion in New York which Professor Einstein has honored with six addresses, and the great appreciation was very well shown by the ovation given to him.

In an interview with a "Campus" reporter, following the taking of the College picture, Professor Einstein remarked, "I am tremendously impressed by your university and by the wonderful men in your College. The honor you showed me by the picture, your hearty applause, struck me with admiration. I was astonished at the close friendship between your faculty and students—a condition rarely seen or even possible in German institutions. It is an admirable example of American democracy. The taking of the picture itself astounded me. To see so many hundreds of young men take their proper places without delay, without remonstrations, without a cumbersome organization is indeed a gratifying sight, truly characteristic of American spirit and ability."

Professor Chase, of the Art Department of this College, who was appointed during the last war to paint the pictures of the great generals, such as Pershing, Foch, Joffre, and others, made three portraits of Professor Einstein.

Much credit is due to Dr. Wetzel, who is well acquainted with Dr. Einstein and has been communicating with him for these several years.

The Campus (CCNY), April 26, pp. 1, 2.

Apparently Professor Alfred N. Goldsmith helped Professor Cohen in putting down Einstein's lectures. "I had the pleasant opportunity of attending your lectures in relativity theory given at the College of the City of New York," Professor Alfred N. Goldsmith writes to Einstein, "at which time I prepared the notes for the series of lectures which notes were later utilized. I shall always remember the clarity and comprehensiveness of your lectures on those occasions."[31] The "utilizer" of his notes was Professor Cohen (see pp. 116–122).

John J. Pershing was commander-in-chief of the American Expeditionary Force in World War I; Joseph J. Joffre and Ferdinand Foch were his French colleagues.

I hope you recognized Franz Liszt's name under the guise of "Lizzt."

Lectures by Professor Einstein

Although Professor Einstein came to the United States primarily in the interests of the Zionist movement, he is giving scientific lectures at various universities. On April 15, he lectured in German on the theory of relativity at Columbia University, under the auspices of the departments of mathematics, physics, astronomy and philosophy, being introduced by Professor Pupin. Professor Einstein was awarded the Barnard medal by Columbia University last year on the recommendation of the National Academy of Sciences.

On April 18, 19, 20 and 21, Professor Einstein gave four lectures before the College of the City of New York on the following subjects: "The 'special' relativity theory;" "Generalized relativity and gravitation;" "The physical significance of entropy and quanta;" "Light-ether and radiation." . . .

Science, April 22 (53 [1921]), p. 382.

Notes on Einstein

We didn't understand a word Prof. Einstein said in chapel, but we never knew it took such a short time to expound his theory.

The next individual who rushes up to us and tells us his original joke about "killing two birds with Einstein" will receive a free ticket to Prof. Einstein's fifth lecture in Doremus Hall.

Prof. Einstein may be considered the greatest living scientist, but don't forget that we passed Physics 3 the second time we took it.

Prof. Einstein was so pleased with the little "Relativity" playlet Gargoyles ran in the last issue of "Campus" that he ordered a hundred copies. He says that it contained the most logical explanation of his theory that was yet published.

The Campus (CCNY), April 26, p. 2.

With Professor Einstein

It is a natural source of gratification to the College that the chief series of lectures given by this celebrated scientist on his first coming to America has been at C.C.N.Y. In four lectures on four successive days he explained his theories and discoveries in detail to such of our staff and friends as could understand his theories and his German. President Mezes, in introducing the lecturer afterward at our General Assembly, appraised him as "among the four or five greatest scientific minds that have ever existed." Of the intellectual power and vibrant fire of the man there can be no question. Our own Professor Morris Cohn, 1900, who has gained repute as an expounder of the Einstein teachings, gives our readers in another column a careful non-technical summary of what Dr. Einstein said in his four lectures here.

The Campus (CCNY), April 29, p. 3.

> Einstein's price went up in three days. On April 26 he was one of the "five or six" greatest scientists of the world, here he is ranked among the "four or five."

A Review of the Einstein Lectures at City College

By Professor Morris R. Cohen, '00

Copyrighted. April 29, 1921. C.C.N.Y.

Prof. Albert Einstein's first appearance before a scientific gathering in this country occurred on Thursday, April 14, when he came to the College to listen to an account of his theory of gravitation by Edward Kasner, '96. Prof. Einstein complimented Kasner for his clear and elegant presentation, and then spoke himself for about twenty minutes on the motives which led him to abandon the Newtonian law of gravitation and to substitute for it his own more general theory, which, though more complicated, has the advantage of explaining phenomena which the Newtonian formula cannot do satisfactorily. It was a rare privilege for the members of the departments of mathematics and physics and the others interested to be thus introduced to the inner motives which actually led to the formulation of that which is certain to be remembered as one of the world's great achievements in science. Particularly illuminating and suggestive was the way Prof. Einstein showed that great and fundamental achievements in science are not brought about by tinkering with details and effecting petty improvements, but rather by radical re-examination of fundamental ideas and trusting the scientific imagination in the suggestion of new departures or lines of development.

I. The Special Theory of Relativity

The first of Prof. Einstein's four public lectures was delivered on Monday, April 18th, before a larger assembly of distinguished physicists, mathematicians and teachers of philosophy from various parts of the country, who were glad to avail themselves of the invitation which the College extended to them, and filled the Doremus Lecture Hall. Prof. Einstein received a most enthusiastic reception when President Mezes appropriately presented him as a scientist of the rank of Newton and Darwin.

With the simplicity of true genius, Prof. Einstein at once began a direct account of the fundamental principle of his original or special theory of relativity. The principle of relativity itself, namely, that all rest or motion is relative to some material body which serves as a center of reference or co-ordinates, is of course an established result of the classical mechanics of Galileo and Newton. The novelty in Einstein's theory is the new conception of time and space, which makes it possible to combine this principle of relativity of mechanics with the principle of the constancy of the velocity of light as fundamental to the modern science of electro-magnetism.

The principle of the absolute constancy of the velocity of light in all directions of space had previously been connected with the conception of a fixed ether, so that this constancy was to hold good only with reference to fixed points in the ether. But the experiments of the American physicists, Michaelson and Morley, showed conclusively that the constancy of the velocity of light in all directions is also true on our moving earth. Experiment thus shows that the principle of relativity is true in optics as well as in mechanics. Hence our notions of time and space, which tell us that this is impossible, that light cannot travel with the same velocity relative to different observers who are moving with reference to each other, must be abandoned. If we accept these, the results of experiment and the workable conventions of science as to the measurement of time, space, and velocity, we must regard our units of time and space and the simultaneity of events as different for different systems that are in relative motion to each other. But that these differences do not conflict with the existence of uniform laws of nature is shown by the Lorentz equations which enable us to calculate the time and space of a moving system if we once know its dimensions when at rest.

II. The Development of the Special Theory

The second lecture dealt with the mathematical development of the original theory of relativity. Prof. Einstein began with the Lorentz equations and went on to show how these equations enable us to examine all the laws of nature and determine whether they are or are not in harmony with the principle of relativity and the principle of the constant velocity of light. He first took up the laws of electric and magnetic fields as formulated by Maxwell and Lorentz and showed how these laws remain true for all systems moving uniformly with reference to each other. He also showed how by means of the Lorentz equations the relation between electric and magnetic fields could be seen in a new light and how it could be demonstrated that electric energy has inertia.

He next took up the law of the conservation of energy and showed how, if we assume that the physical energy of a body is independent of our system of measurement, it follows all energy has inertia, from this it follows that a body loses energy by radiating heat, light or other radiant energy, thereby also loses mass or inertia.

The third or concluding portion of the lecture was devoted to an exposition of the elegant and fruitful mathematical methods of treating the problems of relativity developed by the great German mathematician Minkowski. Professor Einstein brought laughter from his audience by his amusing references to the wide-spread misimpression that the theory of relativity had discovered a fourth dimension of space. He went on to explain that the adherents of the theory of relativity realized perfectly that time could be measured only by clocks and space by yard sticks, and that the two could, therefore, not be identical. But since the theory of relativity shows that the rate at which clocks measure time depends upon their special [spatial] motion, there is great advantage in treating time and space measurements together. In this way the world of physical events can be viewed as a four-dimensional manifold. That means that it is necessary to make four measurements, three space measurements and one time determination, to locate the position of an event. If we understand this we can understand how the Lorentz equations can denote the same mathematical properties that would result from a rotation of our axes of reference in a four-dimensional space.

Before the lecture, Professor Einstein invited questions and comments from his audience. A number of questions were asked which he answered with great patience. In answer to one question he pointed out that the shortening of lengths and the slowing up of clocks were true only on the principle of the constancy of the velocity of light, and that these shortenings held true only if we take our center of measurement in some system with reference to which we are moving. In answer to one inquirer who wished to press him to give the relation of his philosophy to that of Vaihinger's Die Philosophie des Als Ob, Professor Einstein replied that he thought all true philosophical systems could be harmonized with true natural science. To another inquirer he pointed out that while a velocity greater than that of light is conceivable it is not physically possible since it would demand infinite energy.

III. The General Theory of Relativity and Gravitation

For his third lecture, Professor Einstein took as his subject the generalized theory of Relativity. He began with some preliminary remarks about his advanced mathematical methods. But thereafter he stuck rather close to plain prose interspersed with a few geometrical illustrations on the blackboard. The general principle of Relativity means that in detecting motion or describing its laws it ought to be immaterial from what reference body we start our measurement. But if this is so we may well ask why stop at the special theory of Relativity which asserts that the laws of nature are the same in all systems moving uniformly with reference to each other. Why should we not go on to say that the laws of nature must be the same no matter what are the relative motions of the different observers? The usual objection to this is found in Newtonian mechanics which asserts that a body not acted upon by external forces moves uniformly in a straight line, but if acted on by forces it will move with accelerated motion. Now, if two observers are moving relative to each other with non-uniform velocity, the same body which seems to one observer to move in a straight line will seem to the other to move in a curved line, which involves accelerated motion. It would seem, then, that the two observers could not find the same laws of nature governing the motion of the body which they are observing. This objection Professor Einstein showed could be overcome if we recognize the equivalence of mass or inertia and gravity or weight which has been confirmed by experiment with remarkable accuracy. All that we know of gravitation is that in certain portions of space bodies fall or move with uniform acceleration. Thus for an observer in an elevator falling to the earth during a small period of time there is no gravity. Free bodies will not fall toward his floor since the floor itself is falling just as fast. But to an observer on the earth such a free body is subject to gravity and moving with accelerated motion. It is, therefore, always possible, if we know how events would appear in a system that is free from gravity, to calculate how the same events would appear in a system in which they are supposed to be under the influence of gravity. This possibility of passing from one formulation of what has happened to the other proves that both are based on the recognition of the same laws of nature. There are, however, mathematical difficulties in working this out. These difficulties are overcome if we abandon Euclidean Geometry and use instead of the usual system of co-ordinates invented by Descartes, a system of co-ordinates first used by Gauss and an absolute differential calculus developed by Riemann, Christofel, Ricci and Levi Civita. If, in the light of this general principle of Relativity, we re-examine the subject of gravitation, we come to a much more general though a more complicated theory than that of Newton. Newton's theory, however, is seen to be approximately true according to the new theory, so long as we are restricted to ordinary masses and to velocities that are small in comparison with that of light.

If the new theory of gravitation be true there are three facts which are then explained and which can not be explained on the Newtonian theory: first, a certain anomaly in the motion of the planet Mars; second, the bending of light rays that pass near the surface of the Sun, and third, a certain displacement of lines in the spectrum of light coming from the Sun. The last has not yet been satisfactorily determined. Professor Einstein brought repeated applause from his audience by his generous reference to the British scientists who, in spite of the war, fitted out two costly expeditions to test the truth of his theory.

In the concluding portion of his lecture, Professor Einstein dealt briefly with the considerations which lead him to reject the idea of a universe containing a finite amount of matter in an infinite space. An infinite amount of matter seems to be incompatible with the known behavior of bodies. We are, therefore, forced to conclude that space is finite.

The general significance of the theory of Relativity he held to consist in the fact that it shows us the folly of beginning with Euclidean space and then asking how gravity acts in such a world. Instead, we should begin with the study of the way bodies actually behave and formulate the nature of space accordingly.

IV. The Ether and Radiation

The fourth lecture dealt with the ether and the theory of radiation.

The progress of modern electrical theory has enabled us to explain electro-magnetic and optical phenomena by means of the equations of certain fields which hold in a vacuum and do not need any material carrier. The definiteness and simplicity of the resulting laws and the complexity of all the different mechanical models or explanations of the ether has led physicists to drop all the mechanical properties of the ether. In the latest theory, that of Lorentz, the only mechanical property of the ether left, is the negative one of immobility. But on the restricted theory of relativity, the notion of a fixed ether must disappear. For if the ether be at rest with reference to any one system of co-ordinates, it will be in motion in all those systems moving uniformly relative to it. This, however, does not necessarily deny all existence to the ether. It only denies that we can ascribe to the ether any mechanical properties of rest or motion. But in the general theory of relativity, space itself has definite metrical properties in different regions which determine mechanical as well as electro-magnetic phenomena. Space is, therefore, not entirely vacuous and may well be called the ether, provided we do not attach mechanical properties to it and recognize that its properties depend on the presence of matter in it. Einstein thus gets back to a certain extent to Newton's idea of absolute space. Indeed, he expressly justified Newton's use of absolute space to give rotation physical reality (as illustrated in the famous experiment of rotating a bucket of water). On this view, space, or the gravitational ether, is more fundamental than electro-magnetic force. Prof. Einstein, however, prefers to leave open the question of the relation between the ether and the structure of matter. Against those who, like Weyl, would make a premature synthesis of the equations of gravity and electro-dynamics, Einstein protests that the phenomena of radiation may prove insuperable obstacles.

In the second part of his lecture, Prof. Einstein, then, turned to the phenomena of radiation. He commented humorously on the fact that in this field, physicists use on one hand the notion of discrete pulses of energy "pills," and on the other hand radiation that is continuous, "like a soup." Nor do the physicists let the right hand know what the left hand is doing. Not only are they inconsistent, but they fail to explain the facts. Thus, according to the classical view of photo-chemical action, the radiant energy of light breaks up the material molecule. In fact, however, the action will take place no matter how weak the light and will not take place at all, no matter how intense the light, unless this light is of the proper frequency.

Prof. Einstein then sketched Planck's theory of quanta, according to which energy is radiated only in discreet pulses. (Others usually refer to this theory as the Planck-Einstein theory.) This theory avoids logical inconsistency and is more in harmony with the facts. It involves the application of the laws of probability to physical systems; and Prof. Einstein sketched some of the leading ideas in this field which enable us to determine the conditions of equilibrium and the direction in which phenomena must take place.

It was very interesting and instructive to note how Prof. Einstein's treatment of the doctrine of relativity differed from his treatment of the doctrine of quanta. In the former we had a very logical exposition of the fundamental principles of a theory whose main outlines were simple and definitely established in Einstein's own mind. In the treatment of the doctrine of quanta, however, we saw a great mind grappling with problems not yet definitely solved, but determined to try certain modes of attack as most promising. Incidentally, it also illustrated the difference between a physics which proceeds by generalization from experience—by what Rankine called the abstractive method—and a physics which proceeds from hypotheses as to the occult structure of matters in terms of "oscillators," "resonators" and the like. Many have disputed about the relative merits of these two methods. But wise men like Einstein use both.

The achievements of a great man in a definite line of work such as mathematical physics can be adequately appreciated only by experts. But the charm of a great personality is immediately felt by all. Prof. Einstein's personality charmed and impressed everybody. His straightforward simplicity, kindliness, patience, genuine modesty and naive humor almost made one lose sight of the transcendent mental power.

We all have reason to be grateful to Mr. Wetzel, of the Physics Department, for securing the arrangements which brought Prof. Einstein to the College of the City of New York. It is hoped that Prof. Einstein can be induced to publish one of his lectures and thus leave a permanent memorial of a notable event in the history of the College.

The Campus (CCNY), April 29, pp. 3, 4.

Here again another planet is preferred over Mercury: Mars. It must have been the typesetter and not Cohen who thought that Mercury was just a drop of liquid metal, big enough to be capitalized but too liquid to be a planet. Or rather it is the stenographer who should be blamed: note the "special" instead of "spatial" in the text

As other articles listed here show, Kasner's lecture was delivered not on April 14 but on April 8, and Einstein read his first lecture not on April 18 but on April 15.

"At urging of the *New Republic* editors," Cohen writes, "I wrote several more or less popular articles in 1920 and 1921 on the theory of relativity, hoping to use them some day as bricks in the edifice of a popular volume on the foundations of science. When Einstein came to this country and lectured at City College in 1921, I happened to be the most readily available person who understood both his language and his mathematics, and so I was asked to translate his lectures. This gave rise to the altogether undeserved popular legend that I was one of the unbelievably few people in the world who understood the Einstein theory. I suppose that it was because of this legend that *Vanity Fair* paid me one hundred dollars to look at a movie on Einstein's theory of relativity and comment on it. It was the only movie I had ever gone to. I have often expressed a willingness to go to another movie on the same terms, but have found no takers."[32]

Meeting with Einstein made a deep impression on Cohen. As he wrote to Einstein on June 6, 1921, "Poor health, too much drudgery, and a scattering of my mental efforts & the four corners of the intellectual world—law, mathematics, history of civilization, and philosophy—have prevented me from as yet accomplishing any of the things for which I have been striving. The kind and friendly interest which you expressed in my work has, therefore, meant much more encouragement to me than you can readily imagine."[33]

Einstein was also impressed by Cohen. "I knew him well as an extraordinarily helpful, conscientious man of unusually independent character."[34]

On April 23, the Research Committee of the College asked Einstein to put down his last lecture, especially the portion of it on radiation and quantum theory in order to publish it in English, because the quantum theory "is still in need in this country of the stimulus which any publication of yours would give it."[35] This request was not granted.

The *City College Quarterly* remembered Einstein's visits as follows:

"We have lately enjoyed the visits of prominent men to our halls. The one most in the public eye was the famous Professor Einstein, who visited the College no less than six times, being welcomed by the President as a scientist of the rank of Newton and Darwin. He delivered four public lectures during the month of April, before a large assemblage of distinguished physicists, mathematicians and teachers of philosophy from various parts of the country who gladly availed themselves of the invitation which the College extended to them. Since his subjects are of such apparent difficulty to the ordinary lay mind, it may be worth nothing that the *Campus* of April 29 contained a very careful four-column synopsis and review of his lectures by Professor Cohen."

The City College Quarterly, June 2.

Hadassah Entertains Mrs. Weizmann and Mrs. Einstein at Reception

On Thursday afternoon, April 21st, Hadassah tendered a reception and tea to Mrs. Chaim Weizmann and Mrs. Albert Einstein. The Chapters that participated were: New York, Brooklyn, Newark, Plainfield and Bayonne. Miss Lotta Levensohn, National Chairman of Hadassah, presided. Musical numbers were furnished by Mrs. Herman Block and Mr. Davis.

The speakers were: Mrs. Nathan Straus, Mrs. William D. Sperberg, for the Council of Jewish Women; Miss Sarah Kussy for the Women's League of the United Synagogue. Other guests were Mrs. Irving Lehman and Mrs. M. M. Travis.

Arrangements were in charge of Mrs. J. Spiegel, Mrs. Sol Cohen, Mrs. A. Helitzer, Mrs. M. S. Zenker, Mrs. A. Munzer, Mrs. J. Seydel and Mrs. B. Bolrow.

The tables were beautifully decorated with field flowers furnished by Mrs. Nathan Straus.

Mrs. Weizmann spoke of the activities of the Women's International Organization and of the Jewel Fund campaign launched for undertakings in Palestine.

The New Palestine, April 29, p. 4.

Friday, April 22

Meeting Distinguished Men through Their Wives

By Marion Weinstein

If there is anyone as interesting to a woman as a great man, it is the great man's wife. She is the living mirror that reflects his personality, sometimes flatteringly, sometimes relentlessly, as mirrors are wont to do, but always more or less truly. In her eyes and words and bearing, you can catch the reflection of the great man as he is, off life's dress parade.

That is why, in a sense, I was more eager to meet Mrs. Chaim Weizmann and Mrs. Albert Einstein than their distinguished husbands, who are now visiting this country for the first time. And I took leave of these two frank, friendly women in their suites at the Hotel Commodore, with the conviction that the world-renowned scientists whose names they bear are even greater men than they appear to enthusiastic American Jewry.

Despite the fact that Mrs. Weizmann and Mrs. Einstein are two of the mostmarried wives I have ever seen, they have retained all the color of their own distinct individualities. What that fact bespeaks for the professors, ask any wren-like wife of a great man whose personality has been drained to a shadow. Both radiate the calm, sweet self-assurance of women who know daily tenderness. In the eyes of both—and they are too honest and too intelligent to deceive either themselves or the close observer—their husbands sit on the loftiest pedestals, and, I am told, it is even harder for a man to remain a hero to his own wife than to the proverbial valet.

Aside from their striking likeness as obviously happy wives, these women are two distinctly different types, as are the men whose lives they share, though both historic figures in the field of science and the Zionist movement. Dr. Weizmann is dynamic, self-assertive, life-loving; Professor Einstein is the serene dreamer, self-effacing, a lover of nature in her gentler moods and of the simple joys of life.

Mrs. Chaim Weizmann, a pretty, vivacious brunette, almost girlish in her chic French frock, gives the impression of the cultured Parisienne. She is highly ornamental to any scene and yet has uncommon gifts of mind and character. Her conversation ripples and sparkles in fluent English whose charm is enhanced by the faint suggestion of a Slavic flavor. She confesses, with the most winning of smiles, that she loves pretty clothes and the good things of life. She registers what is known among connoisseurs, social and commercial, as "expensive simplicity." It is particularly striking that she does not wear a single jewel.

Mrs. Albert Einstein, on the other hand, had the beauty of a fine painting. She is the type men love for the innate womanliness, artists choose for their "Madonnas" and Jews picture as the ideal "mother of Israel." You associate her with deep, refreshing pools and balmy sunshine. Her simplest replies reveal a deeply contemplative nature. And yet, with all her stability, she has a rollicking sense of fun. Her fine blue eyes and pretty teeth flash with laughter, as she tells you some of her *amusing* experiences in this maelstrom of America.

Speaks a Quaint English

She speaks a quaint but intelligible English, of which the Professor knows but few words. She is his constant companion, his business manager and general "bodyguard." Even while I was calling he had occasion to consult her twice. And his "Elsa" had the information he sought at her fingers' ends.

America certainly presents unique trials to Mrs. Einstein. Never has the Professor been besieged by so many persistent pursuers, who appear at all hours, to his exhaustion. They crowd the corridors, they clamor at his door night and day, particularly the ingenious publicity people. One day the Professor, driven to desperation, took matters into his own hands and disappeared for a few hours. A mad search began for him. Even Mrs. Einstein felt a faint twinge of alarm. When every effort to find him had failed, he suddenly stood before her. In one hand he clasped his beloved violin and in the other his inseparable briar pipe. He had been driven to lock himself, he explained, in his "Bade Zimmer."

Like Mrs. Weizmann, Mrs. Einstein has had a career of her own. Before her marriage to Professor Einstein, three years ago, she was a widow with two daughters giving lectures in Berlin in modern and classic literature. She was but eighteen at the time of her first marriage and is a cousin of the Professor on both sides of the family. Her husband has two sons from a former marriage, Edward and Albert Jr., still schoolboys, whose photograph she shows with great pride.

In the early morning hours when I saw her, Mrs. Einstein was wearing a simple cloth skirt and a purple silk, daintily embroidered overblouse. I asked her if she liked fine clothes.

"I do like them," she answered, "but in Germany we have grown unaccustomed to them since the war. They are very difficult to get and extremely expensive."

Her greatest pleasure is to walk with her husband among the beautiful byways of nature. Professor Einstein loves particularly to stroll along the level stretches that abound about Berlin. Although he has spent so many years in Switzerland, among the glories of the Alps, he does not like high mountain vistas. Both are fond of hours with old friends, but do not care for society. They share, too, a love of music. Professor Einstein is a proficient violinist and his wife used to excel on the piano. However, nothing delights the professor more, she says, than to get aboard the yacht of some dear friend and drift on the German waters. The sport of yachting has no charm for him. What appeals him is the sheer aesthetic pleasure of sailing.

Mrs. Einstein's interest in Zionism dates back to two years ago when the professor, stirred by the plight of hundreds of Jewish students in Berlin, who were denied entrance to the universities, became a Zionist. Her great Zionist enthusiasm, as in the case of the author of the theory of relativity, is the future Hebrew university, whose cornerstone was laid three years ago on Mount Scopus in Jerusalem.

Mrs. Weizmann has been a Zionist, on the other hand, for eighteen years, ever since she joined a Jewish nationalist club as a student in Geneva. Russian-born, she holds degrees in medicine from the Universities of Geneva and Manchester. She was a practicing physician in Manchester and municipal medical officer in charge of seven baby clinics there, when she married Dr. Weizmann fourteen and a half years ago. They have two sons, Benjamin Isaiah, thirteen years old, and Michael Oser aged four.

Both Moderate Feminists

Like Mrs. Einstein, she is a believer in suffrage and opportunities for women. Both, however, are moderate feminists. "I do no think the vote is of the greatest importance," says Mrs. Einstein. "I do not believe in women trying to be men," says Mrs. Weizmann. "Suffrage would then only add numbers. I still remain a woman."

Although she and Dr. Weizmann are great comrades and he consults her at every turn, she has never looked very deeply into his activities as a master chemist. It is his labors as the world leader of Zionism that she has taken the greatest interest, an interest that has grown even greater since she accompanied him to Palestine eighteen months ago. Her active contribution to the Zionist cause is in connection with the women's work for women in Palestine. Just now this is taking the form of the Jewel Fund.

"The Jewel Fund was inaugurated," Mrs. Weizmann told me, "by the Women's International Zionist Organization, which was itself formed at the Women's Zionist Conference in London last July. It is part of the Keren Hayesod. It is our aim to appeal to all Jewish women the world over for jewels and money, the proceeds to be devoted to our new schemes for women's welfare in Palestine that have been approved by the Zionist Organization.

"Lady Herbert Samuel is president of the Jewel Fund and I am the treasurer. As our schemes are of a purely humanitarian character in Palestine we feel we can appeal to all Jewish women, Zionist and non-Zionist, for support. We have secured on our Committee a representative of the Union of Jewish Women, the largest women's organization in Great Britain. Mrs. Herz, the wife of the Chief Rabbi, is another member and so is Mrs. Rothschild Behrens, the sister of Lord Walter Rothschild, as well as many other prominent women."

The work for the Jewel Fund has been carried on through a series of drawing-room meetings in all the districts of London. Because the value of jewels has fallen, the refugee Russians having heaped so many upon the European market, the women have sought to raise funds, too, through concerts, lectures, and bridge parties.

The first drawing-room meeting was held at the home of Mrs. Weizmann, Lady Samuel presiding. At this and subsequent gatherings, enthusiasm ran high. There was many a touching little incident as women parted with valued heirlooms and keepsakes. The Committee had asked them to give not only what they could spare, but what they cherished, so that they might feel they were making a distinct personal contribution to the upbuilding of Palestine. A girl none too prosperously dressed, would rush up with her engagement ring; another with a modest wrist watch. One woman sent Mrs. Weizmann anonymously her little baby's ring. Poor women with no jewels to give sent bits of broken gold and old silver or coins they could only spare with sacrifice. Mrs. James de Rothschild, the wife of the Major who recently visited America in the interests of Palestine, sent to Mrs. Weizmann one of her most magnificent wedding presents, a wide diamond bracelet with three emeralds.

Both Mrs. Weizmann and Mrs. Einstein find it a great burden to be the wives of illustrious men. "You have to share your husband with so many interests," says Mrs. Weizmann, with a pretty pout. "Your husband is public property," sighs Mrs. Einstein. But both agree emphatically that the disadvantages are far eclipsed by the great compensations.

The American Hebrew, April 22, p. 632–33.

Einstein Wrong, Brush Indicates

New Experiments in Gravity Startle the American Philosophical Society— Variation of Force Found—Inventor Declares Theory Behind Results Upsets Views of the Earth's Density

Special to *The New York Times.*

PHILADELPHIA, April 22.—Perhaps Dr. Einstein is wrong after all.

Dr. Charles F. Brush of Cleveland, inventor of the Brush electric arc light, the electric storage battery and many other electrical devices, described experiments in gravitation to the American Philosophical Society today which, members of the society declared, proved, if accurate, that the Einstein theory was incorrect....

"If these experiments are accurate," commented Dr. A. G. Webster of Clark University, Worcester, Mass., "they absolutely disprove the Einstein theory."

Dr. Brush explained that he had taken particles of bismuth and other metals of the same "units of mass" and attached them to pendulums. Watching the pendulums in their swing led him to conclude that "the earth gravitational field grips bismuth stronger per unit of mass than it grips zinc per unit of mass. Bismuth weighs more per unit of mass than zinc."...

Another scientist, discussing the paper, said that if these experiments did upset the Einstein theory, no one would be better pleased than Professor Einstein himself, for he had said recently in conversation that he wished some one would either prove or disprove his theory, so that further progress could be made.

The New York Times, April 23, p. 1.

Einstein was invited to address the annual general meeting of the American Philosophical Society by its president, W. B. Scott.[36] The invitation was also forwarded in person by Oscar S. Straus at a dinner with Einstein on April 5. Einstein "makes a charming impression as a plain, unassuming, modest and dreamy philosopher."[37] A dreamy philosopher or just a confused physicist? Scott thought it better to put some pressure on him, so he sent Professor Goodspeed to visit him in person "as I think that is quite indispensable.... Veblen and Eisenhart report him as somewhat overwhelmed and confused with his engagements."[38] All the same, on April 15 Einstein regretfully admitted that he would be unable to attend the meeting.[39]

You will find details about Brush's views on p. 189; here I present only excerpts from the article.

G. H. Campbell, A. N. Goldsmith, W. A. Winterbottom, Einstein, C. P. Steinmetz, and D. Sarnoff with apparatus for transatlantic radio communication in New Brunswick, April 23. Courtesy Bildarchiv Preußischer Kulturbesitz, Berlin.

Saturday, April 23

Einstein Sees Radio Tests

Witnesses Speed Demonstration at New Jersey Station

NEW BRUNSWICK, N. J., April 23.—A special demonstration of high speed transatlantic wireless radio transmission was given at the New Brunswick station of the Radio Corporation of America this afternoon for the benefit of Professor Albert Einstein, the scientist.

Professor Alfred Goldsmith, head of the Radio Engineering Department of the College of the City of New York, and Dr. C. P. Steinmetz, consulting engineer of the General Electric Company, accompanied Professor Einstein. Messages were sent at the rate of sixty words a minute to four European countries.

Signals were shown on the local oscillographs showing the effect of magnetic amplifiers and radiation from antennae, and the messages were also followed by sound signals demonstrating the advisability of high speed transmission for regular commercial work.

Communication was held with ships at sea within 500 miles of the American coast. While the visitors were still witnessing the experiments Thomas J. Hayden, in charge of the station, received a cable message informing him that the signals could be heard more clearly in Japan than those sent out from stations on the Pacific Coast.

The New York Times, April 24, p. 6.

This was not the first demonstration at New Brunswick of a long-range broadcast. The station transmitted news on war preparations in the United States, on raising huge amounts of money for the war, on transporting American troops to Europe—all for the express purpose of being listened to by German stations. "It was an offensive against morale, a psychological attack," John Hammond adds.[40]

Even after the war, President Wilson's Fourteen Points were broadcast to Germany from there, making the armistice negotiations shorter by months. It was rather evident that the political importance of the station was also supposed to be of interest to the pacifist Einstein.

The physicist Einstein also could find some cutting edge technology there: the so-called Alexanderson system of transatlantic telecommunication.[41]

10

Puzzles Harding

✦

Monday, April 25

Einstein's Idea Puzzles Harding, He Admits As Scientist Calls

WASHINGTON, April 25.—The theory of relativity of matter, which got into a Senate debate recently, with assertions by Senators Boise Penrose, John Sharp Williams and others that they did not understand it, has also vanquished President Harding.

Its originator, Dr. Albert Einstein, called on Mr. Harding today with a delegation from the National Academy of Sciences. As the group posed before a camera the President smilingly confessed that he, too, failed to grasp the relativity idea.

Dr. Einstein is to address the Academy tomorrow at its annual session.

The New York Times, April 26, p. 1.

Tuesday, April 26

Urges Scientists' Reunion

Einstein Deplores the Losses to Learning Attributable to War

WASHINGTON, April 26.—Deploring the losses to science "through the action of political misfortune," Professor Albert Einstein, propounder of the theory of relativity, speaking in German before the National Academy of Sciences today, expressed the hope "that the field of activity of scientific men may be reunited and that the whole world will soon again be bound together by common work."

"Prof. Einstein, Famous Swiss Scientist Who Originated Theory of Relativity, Calls on Harding. Left to Right.—Mrs. Albert Einstein, Prof. Albert Einstein, President Harding and Dr. Charles D. Walcott, secretary of the Smithsonian Institution."
The Washington Post, April 26, p. 2.

An almost identical photograph was published in the *New York Times* Rotogravure Section, May 1, with the following caption: "Is He Thinking of Relativity? President Harding, Who Admits He Is Not One of the Twelve Who Understand the Theories Advanced by Professor Albert Einstein, Receives Their Author at the White House, with Dr. Charles D. Walcott, President of the National Academy of Science, Whose Guest Professor Einstein Was." No mistake: Walcott was both secretary of the Smithsonian and president of the Academy.

Courtesy Corbis.

The eminent Swiss scientist, in acknowledging a high tribute paid him by President Charles D. Walcott of the Academy, said the appreciation embarrassed him.

"When a man after long years of searching," he said, "chances upon a thought which discloses something of the beauty of this mysterious universe, he should not therefore be personally celebrated. He is already sufficiently paid by his experience of seeking and finding. In science, moreover, the work of the individual is so bound up with that of his scientific predecessors and contemporaries that it appears almost as an impersonal product of his generation."

President Walcott in greeting Professor Einstein said:

"The Academy rejoices to bring its tribute of homage to the brilliant and penetrating mind which has so greatly enriched the philosophy of ultimate truth. We congratulate you on the universal appreciation of your investigations, which have outrun and overleaped the limitations and barriers associated with nationalities and with the times. To men everywhere your name, in association with the abstruse subject of your investigations, has become a household word."

The New York Times, April 27, p. 21.

President Charles D. Walcott invited Einstein to the meetings of the Academy on April 18.[42] He added that Gano Dunn, Columbia University professor, member of the National Academy of Sciences, and person "in charge" of taking care of the Einsteins in Washington, informed Einstein personally that a brief address of welcome would be presented to him on Tuesday, at about 12:30, just prior to the adjournment of the morning session. Aware of the fact that Einstein did not speak English "fluently" (actually at the time he did not speak English at all), Walcott sent him the German translation of the text of what he would say on behalf of the Academy to help Einstein in formulating his response. Walcott also expressed his thanks for Einstein's having given consent to address the Academy at the dinner. The greeting sent to Einstein follows the text published in the *New York Times* word for word but adds the following sentence: "We welcome you to our scientific meetings and especially to the social hours which intervene, during which the members of the Academy hope to have the pleasure of meeting and learning to know you as a friend."

WHEN PROFESSOR EINSTEIN CALLED ON PRESIDENT HARDING

Harding, still a senator but presidential candidate, was very careful about his use of the word "normalcy":

"I have noticed that word caused considerable newspaper editors to change it to 'normality.' I have looked for 'normality' in my dictionary, and I do not find it there. 'Normalcy,' however, I did find, and it is a good word.

"And here is the definition:

"By 'normalcy' I do not mean the old order, but a regular, steady order of things. I mean normal procedure, the natural way, without excess. I don't believe the old order can or should come back, but we must have normal order, or, as I have said, 'normalcy.'"

The New York Times, July 21, 1920, p. 7.

As clear as day.

The [New York] *Call Magazine,* May 1, p. 8.

Walcott was the second American to propose Einstein for the Nobel Prize.[43]

The ambiguity of Einstein's citizenship eclipsed Einstein's reception in Washington. "There was general cordiality but aloofness in certain directions owing to the misunderstanding that Dr. and Mrs. Einstein were German citizens," Gano Dunn writes. "I explained this away in several instances and it made a big difference (I am rather sorry to say) although I can fully understand and actually share the feeling against Germans, including German scientific men who left their field of science to sign the round robin, etc. Einstein while born of German parents, is a Swiss citizen who during the war accepted the call to leave Zurich and teach at Berlin only on the written agreement of the Emperor that he should maintain his Swiss citizenship."[44] See pp. 64–65 for an explanation of Einstein's citizenship status.

The "round robin" was the Manifesto described on p. 215.

Scientists Receive Medals of Honor

Prince of Monaco, Admiral Sigsbee and Dr. Ridgway Among Recipients—Measures Stars by Light—Dr. W. S. Adams Announces Discovery—Prof. Einstein Deplores Injury to Science by World War

(By the Associated Press.)

The Alexander Agassiz gold medal for original contributions to the science of oceanography was presented to Prince Albert I of Monaco, at the annual dinner of the National Academy of Sciences last night. The Henry Draper gold medal for eminence in investigations in astronomical physics was presented to Dr. P. Zeeman, of Holland, in recognition particularly of his researches on the influences of magnetism on light. . . .

Dr. Robert Ridgway, of the National Museum in this city, was awarded the Daniel Giraud Elliott gold medal for his studies of the birds of North America, and Dr. C. W. Stiles, of Wilmington, N.C., was given the gold medal for eminence in the application of science to public welfare, in recognition of his studies of the hookworm.

Through analysis of star spectrums, Dr. W. S. Adams, of Pasadena, Calif., announced at the day session of the academy that he had determined . . . that the position of stars as well as their direction and rate of movement could be accurately estimated. . . .

Deploring the impairment to science "through the action of political misfortune," Prof. Albert Einstein, propounder of the theory of relativity, speaking in German at the day session, expressed the hope "that the field of activity of scientific men may be reunited and that the whole world will soon again be bound together by common work."

Advantages of the Old Sol cook stove, operated 24 hours a day on sun heat alone, were explained to the National Academy of Sciences here yesterday by Dr. C. G. Abbott, of the Smithsonian Institution. . . .

The Washington Post, April 27, p. 3.

On the annual dinner that followed the session of the Academy, "we talked and gossiped," writes Harlow Shapley, the famous Caltech astronomer. "Then came the speech making. It was just horrible. The Academy gives various prizes and hands out medals of one kind and another. On this occasion it was honoring the Prince of Monaco for oceanography, and several others. One I particularly remember was some noble operator from a government bureau who had worked successfully on hookworm control. That was his specialty, his life.

"But none of these people were exciting as speakers. They did not know how to make a talk. One would talk for a long while about 'Johnson the Scientist,' 'Johnson as Operator,' 'Johnson the Man'— those were the kinds of phrases they used. And then another would get up and do much the same. I groaned.

"Over at the end of the head table there was a visitor from Europe named Albert Einstein. He was sitting there with a beaming face that reflected his wonderful personality. Alongside him, I believe, was the secretary from the Netherlands embassy who was there on behalf of the physicist Pieter Zeeman to receive a prize, a medal, or something of that kind. And here all these non-speakers droned on and on. To me it was embarrassing, to others also, but Einstein, smiling, leaned over to the Dutchman and whispered something. The Dutchman turned away quickly to hide his big smile or possibly guffaw. 'What did Einstein say' we asked the Dutchman afterward. 'He said, "I have just got a new theory of Eternity!"'"[45]

During this dinner, Gano Dunn, on the request of Oswald Veblen of Princeton, asked Einstein not to limit scientific and technical material to the last three of his planned Princeton lectures—a request that was only partially fulfilled. In his Friday letter, he gave Veblen "instructions" how to take care of the Einsteins: "I discovered in taking the Einsteins around and in seeing them in New York . . . that one of the things they particularly appreciate is having somebody with them all the time to relieve them of the burden of finding out where to go and how to get there. If somebody's motor and driver, together with some representatives of the faculty as a guide, could be on hand at all times to tell them what to do and take them to and fro, it is the one thing I think they would most appreciate. This was my job at Washington and they have never stopped talking about it. Dr.

Einstein speaks practically no English, although some French. Mrs. Einstein speaks English but does not always understand and is fearful lest by misapprehension she will expose her husband to embarrassment. Wherever they stay, the photographers and newspaper men almost break down the door, although this of course will not happen in Princeton. If they stay at the hotel, they should be protected as far as possible, so that they can have some time each day to themselves and not get all tired out."[46]

The same day L. D. Brandeis first met with the Einsteins at 6 p.m. They spent an hour together. As Brandeis confessed to Stephen S. Wise, "I struggled in German to make them understand the situation as they are eager to talk on it." They concluded that Einstein will confine himself to the matters of the university. "I think it would be advisable," Brandeis continued, "to talk with them—not in the presence of W[eizmann]."[47]

11

Speaks for Proposed Zionist University

✦

Wednesday, April 27

Einstein and Weizmann to Address Students

Prof. Albert Einstein and Dr. Chaim Weizmann will be the guests and speakers at a special meeting of the Council of Jewish Student Societies, to be held in the Great Hall of the College of the City of New York, this evening at 8 o'clock. Both distinguished scientists will speak on "The Proposed Hebrew University in Palestine," in behalf of which Professor Einstein and his colleagues are visiting this country. The Hon. Oscar S. Straus, former United States Ambassador to Turkey, will preside. The meeting will be open to students, graduates and all others interested in the proposed Hebrew University.

The Council of Jewish Student Societies consists of representatives of national organizations of Jewish students, including the Intercollegiate Menorah Association, the Intercollegiate Zionist Association, the Zeta Beta Tau, Beta Sigma Rho, Alpha Epsilon Pi, Tau Epsilon Phi, Kappa Nu, and Phi Epsilon Pi Fraternities.

Yidishes Togeblat—The Jewish Daily News, April 27.

Einstein to Students

With Dr. Weizmann He Speaks for Proposed Zionist University

Professor Albert Einstein and Dr. Chaim Weizmann, President of the World Zionist Organization, spoke last night at a meeting of the Council of Jewish Student Societies in the hall of the College of the City of New York. Nearly

2,000 persons were in the great hall to hear the two leading Zionists speak in behalf of the proposed Hebrew University in Palestine.

They were introduced by Oscar S. Straus, former Ambassador to Turkey, who told of the achievements of the two men and told briefly of the work they are doing in Palestine.

The New York Times, April 28, p. 14.

Thursday, April 28

On this day, Einstein asked Louis D. Brandeis for certain data, which Brandeis sent him on the following day via Ben Flexner.[48] One of the bones of contention between Weizmann and Brandeis was whether the funds collected in the United States for the university had been spent for other purposes.[49] It appears that Einstein had wanted clarity in this matter.

Friday and Saturday, April 29–30

Einstein to Lecture Here Tomorrow

Professor Einstein, exponent of the theory of relativity, will deliver two lectures on the "Cardinal Principles and Methods of the Theory of Relativity," before the Zionist Society of Engineering and Agriculture at the Engineers' Building, 29 West 39th street.

The first lecture will be given tomorrow evening, and will be followed with the second on Saturday.

The New York Call, April 28, p. 6.

Einstein was invited to lecture by Executive Secretary Isaac J. Stander on April 25.[50] The lectures were to be delivered at 8:15 p.m. at the Auditorium of the Engineering Societies Building. There are no reports on their reception.

Sunday, May 1

Einstein on Irrelevances

By Don Arnald

"How comfortable you make everything in the hotel! Every door, every window, is perfect; nothing is out of order. It is all so well planned and well organized. I never saw such rooms; such care for details; such hotel lobbies, with so many to serve you. Everything—everything is systematized, down to the bathrooms. You people in America are very practical. I like the way you light up the windows with the signs. I like the cheerful way you arrange the electricity up and down the streets."

So spoke Professor Albert Einstein, apostle of relativity, in the course of a talk about his experiences in New York.

"What was it that impressed you most when you arrived?" the interviewer asked.

"Ah! I see so many nationalities living together so well. America is a country of many different peoples at peace with one another. Then, too, I like the restaurants with the 'color' of the nations in the air. Each has its own atmosphere. It is like a zoological garden of nationalities, when you go from one to the other."

"Are you a bit disappointed not to find some beer in our dining rooms?"

"I cannot say alcohol is as bad as people think it is," replied the professor. "It may not be so good for men to spend all their wages on drinking. But it is more an economic question than a question of health. Some workmen must have liquor, it seems. We must not take everything away. Prohibition shows the strength of your democratic Government against private interests. In a corrupt State this could not be done."

"Do you consider it against personal liberty to take liquor away?"

"How could that be in America? You have a republic. You have no dictator who makes slaves of people. Nothing that is done by a democratic Government could be done against freedom. I think you will find it best for the economic welfare of the people in the end."

"How about tobacco?" was the next question. "Some people want to take that away, too."

Prof. Einstein drew back in surprise. "Oh my, no! I never heard of it. So some one is starting this? Who is doing this?"

"Some temperance organization here in the United States."

The professor said: "If I do not wish to smoke, I say it is excellent to take my tobacco away. But I do wish to smoke, so I say I do not like you to do that."

"But they say it is not healthful."

"If you take our tobacco and everything else away, what have you left?" cried Professor Einstein. "It may be healthful to take away tobacco, but it is mighty lonesome." He thought a moment. "But this is economic, too," he said, at last. "The men spend too much money on cigars, and their wives kick; therefore, they take it away. They say it costs too much money to smoke. I do not know! I have never heard of such a thing as taking away a man's smoking! I'll stick to my pipe. I do not care who will not smoke. I will! If you take everything away, life is not worth while!"

"And the blue laws—how about them?"

"Blue laws? Blue laws? I never heard of those blue laws in my life. What are you saying?" The professor fairly blazed with consternation.

"They want to pass laws to close up all places of amusement on Sunday," the interviewer explained. "All theatres, music shows, baseball and other places will be shut down, including everything for relaxation, even amusement parks and the movies."

"For Heaven's sake. More laws? I have never heard of such a thing. Here's what I say: Men must have rest, yes? But what is the right rest? You cannot make a law to tell people how to do it. See—some people rest when they lie down and go to sleep. Others have rest when they are wide awake and are stimulated. They must work or write or go to amusements to find rest. If you pass one law to show all people how to rest, that means you make everybody alike. But everybody is not alike. No, I do not care for these blue laws. They will do no good for the country or the people.

"Many workmen want to go to movies on Sunday because they have no time during the week days, so they find rest there," he continued. "And that is very good."

"What do you think of our movies and the theatres?"

"I've been so busy that I haven't had much time, but I have never in my life seen such theatres—everything for your taste, all sorts of plays, comedy, tragedy, romance, pageants. And the movies? I am enthusiastic about them—I mean for the presentation of living moving things. They will develop more and more. In general, the pictures shown now are not so artistic, but they will get better, very much better, all the time. The art is not high enough now, but soon you will have science through this art, as well as you are now having art through this science. I see how the movies will be used in the future for science in bacteriology and technology. Perhaps not so soon for astronomy, because the motions of the heavenly bodies are too quick for measurement. But the movies must only be fitted well, and they can be used most adequately for instruction in all science! I think, all in all, the movies are only in their infancy. They are very beautiful, but they will get better, until the best plays can be shown. You deserve much credit for doing such fine pictures. I compliment you, and I hope for more artistic plays right along."

At this point his wife, a charming little gray-haired lady, slipped into the room and sat by her husband's side.

"Maybe I can help you," she said kindly. "I speak English, and I can interpret for him." The interview up to that point had been in German.

"Perhaps you can tell me something about the professor's life," I asked. Dr. Einstein laughed heartily.

"He does not want my life," said he. "That is of no use to him. Why should he care for that. He is asking what I think of New York. I tell him glorious! I tell him I see here the greatest city in the world, like Paris, like London, only better! I tell him here all people of all nationalities are melted together—and are happy. I tell him the stranger comes here and is full of joy because he goes to his people at once and feels at home."

"But your book on relativity translated into English, maybe he wants that," queried Mrs. Einstein.

"No; why that!" said the professor. "He doesn't come here for relativity. He comes here to see me. I want to say something to the people, how I like the restaurants and the theatres and the movies and the hotels, and how I do not like the blue laws—and if they take away my tobacco—I do not know what I'll do, but I'll take America anyway, no matter what they do."

At this the secretary arrived. He wanted to add a word on the professor's mission in America. He said:

"I suppose you know Professor Einstein is here to help the University of Palestine. Its foundation stone was laid by Dr. Weizmann in 1918, and since then the university site has been expanded. There is also a library with more that 3,000 volumes and rapidly growing. Plans have been worked out both for the complete university of the future and for a comparatively modest beginning. The time has now come for us to make a foundation fund, part of which will go to the university. American people play a great part in world politics, showing that their aspirations are noble, and we have come from sick and suffering Europe with feelings of hope, convinced that our spiritual aims will command the full sympathy of the American Nation."

Dr. Einstein broke in: "We will receive their enthusiastic approval, we are sure, but the people know all this. This gentleman asks me other things, and I tell him what I think of New York."

He slapped me on the back and added: "You greet for me all the good people of America and you say, 'I feel at home here among people, many different people from all the nations in the world.'"

The New York Times Book Review and Magazine, May 1, p. iii.

> If Einstein had seen that "politicians are ducking, candidates are hedging, the Anti-Saloon League prospering, people are being poisoned, bootleggers are being enriched, and Government officials are being corrupted," he would not have made these complimentary remarks. The quote is from the pen of future mayor of New York Fiorello La-Guardia, an opponent of the Prohibition Laws from the outset and president of the Council of Aldermen of New York in the spring of 1921.[51] As you saw on p. 54, he had the task of paving the way to Weizmann and Einstein getting the privilege of freedom of New York City.
>
> Mr. Ginzberg here talks about 3,000 volumes in the library of the nascent Hebrew University, but again the typesetter may have erred, probably missing a zero here; other accounts mention 30,000 volumes. Notice also how Ginzberg tries to get Einstein not to talk too much about himself but rather about the University!

12

Baffled in Chicago

✦

Monday, May 2

Dr. Einstein Here Today to Explain His "Relativity"

Just what is the theory of relativity? At least one man knows—Dr. Albert Einstein, "the man who made it"—and he will be here today. University of Chicago and Northwestern university professors and other scientists of the city, many of whom at least privately confess to some uncertainty in their knowledge of the celebrated theory, will sit at Dr. Einstein's feet through three lectures in Mandel hall, University of Chicago. The lectures, each starting at 4:30 o'clock and lasting one hour, will be given tomorrow afternoon, Wednesday and Thursday.

Dr. Einstein, who is in Chicago in the interest of a drive for funds for the endowment of the Hebrew university in Jerusalem, now being erected on the Mount of Olives, maintains his conception of relativity cannot be properly expounded in less than three hours, and has steadfastly refused to undertake treatment of the subject in less time.

The famous scientist will be the guest at a reception tonight at the home of Max Epstein, 4906 Greenwood avenue.

Chicago Daily Tribune, May 2, p. 7.

> There is not much available information on Max Epstein, except that he was a business executive and a philanthropist.

Einstein Here, Not Relatively, But in Flesh

Says He Needs Five Days to Expound Theory

The *Tribune* regrets to inform its readers that it will be unable to present to them Prof. Einstein's theory of relativity. Yesterday a reporter was sent to interview Prof. Einstein and get a column story on the theory. After the professor, through an interpreter, had explained that the most incidental discussion of the question would take from three to four hours, and that a thorough discourse might be completed in five days, it was decided to confine the interview to other things.

Prof. Albert Einstein, scientist and author of the theory of relativity, which has given the world a new scientific conception of space and time, arrived in Chicago yesterday to lecture on his theory at the University of Chicago. As he will give only three lectures here, however, he said he could not go deeply into the subject, expecting to do that at Princeton university, where he is to talk every day next week.

His Theory? Simple!

When asked regarding the statement that there were only twelve persons in the United States whose mentality was sufficient to grasp the principles of his theory, Prof. Einstein laughed.

"Everywhere I go some one asks me that question," he said. "It is absurd. Any one who has had sufficient training in science can readily understand the theory. There is nothing amazing or mysterious about it. It is simple to minds trained along that line and there are many such in the United States."

The primary object of his visit to America, he said, was in the interest of the proposed Hebrew University of Jerusalem, but he has taken advantage of the opportunity to get in touch with the American scientific world. He wished particularly to meet Prof. Michaelson of the University of Chicago and was sorry to learn he was in Europe. Prof. Michaelson's experiments have done much to prove the truth of his theory, Prof. Einstein said.

Seeks Help for School

"The thing that led me to take such an active interest in the proposed university at Jerusalem," said Prof. Einstein, "was the situation that developed among Jewish students in central Europe after the war. A great number of young Jewish students had completed their preparatory courses, and were ready to enter colleges and universities, but were prohibited from doing so because the schools were overcrowded. Prejudice against my race also has something to do with the situation. As a result, it was difficult for a young Jew to complete his education.

"The ultimate object of my visit is to raise funds with which to promote this great undertaking; but because of the shortness of the time I will spend in this country, I do not expect to raise the money myself. My mission is to arouse sympathy for the plan."

Prof. Einstein said he attached much importance to the establishment of the university, not only as a national Jewish institution, but also as a seat of Jewish culture.

The proposed institution will be coeducational. It already has a library of 30,000 volumes.

Following his lectures at the University of Chicago, which will be given this afternoon, tomorrow, and Thursday, Prof. Einstein will return to New York. Last night he was the guest of honor at an informal reception at the home of Max Epstein, where he met many of Chicago's scientists, scholars, and leaders.

Chicago Daily Tribune, May 3, sect. 1, p. 3.

World's Smartest Man?

Some folks say so. For he is none other than Prof. Albert Einstein, whose theory of relativity has upset some of the most cherished of scientific perceptions. With Prof. Einstein is his wife. They came to Chicago yesterday to remain for several days while the doctor delivers three lectures at the University of Chicago and furthers the cause of a Jewish university in Palestine.

Chicago Daily Tribune, May 3, sect. 1, p. 3.

Einstein Baffled in Chicago

Seeks Pants in Only Three Dimensions—Faces Relativity of Trousers

By Charles Mac Arthur

Before going into this theory of relativity, it is necessary to explain that Professor Albert Einstein, who worked it out, is our guest for a few days at the Auditorium Hotel.

Next, don't expect much. White paper still costs a lot of money and the professor said it would take a week of conversation to give the average listener the faintest idea of what it's all about.

And in the few moments Dr. Einstein had to spare last night there were things far more important than stellar space and metaphysical theorems on his mind.

The Relativity of Trousers

For one thing, there were his pants. At 8 o'clock he was to appear at a formal reception in his honor. At 7:30 his trunks, containing his dress trousers, had not arrived. His traveling pants had not been ironed.

To give the problem its full relative significance:

Supposing the dray containing his trunk, and therefore his pants, to be steaming westward from the railroad depot; supposing the earth to be whirling westward at the same time; supposing the earth to be revolving about the sun in still another orbit; and the solar system to be moving another way—in what direction would the professor's pants be traveling?

It becomes apparent at once that the simplest incident may be crowded with the most abstruse problems to a man of science; and the learned doctor reflected this concern as he paced up and down and answered the childish questions of his interviewers, particularly as they asked him to tell them in five minutes what it had taken a lifetime to discover.

The Very Cross Examiner

"Doctor," said a reporter, "your theory is that space is not necessarily infinite."

Mr. Solomon Ginsberg, his interpreter, nodded.

"It can be bounded."

Another nod.

"If there were a globe trillions of miles in diameter, hollow in the center, could it inclose the entire universe?"

Mr. Ginsberg indicated this was so.

This was the reporter's chance.

"What would be outside the globe?" he asked.

The question was translated with a smile. Both Dr. Einstein and Mr. Ginsberg looked upon the reporter with kindness, mingled with pity.

In Words of One Syllable

"You have imagined your globe," said Mr. Ginsberg. "Now imagine that its surface is covered with organisms of two dimensions. They have width and length, no thickness at all. They cover the globe like cracked paint, each flake, except for its thickness, representing an organism.

"Now, supposing this organism to be able to move around the globe. It could travel for millions of years and would always return to its starting point. It would never be conscious of what was above it or beneath it.

"This hypothetical organism would be oblivious to anything above or below the narrow plane in which it exists."

Roll Your Own Universe

When the reporter came to he was vainly trying to light a three dimensional cigaret with a three dimensional match. It began to trickle into his brain that the two dimensional organism referred to was himself and so far from being the thirteenth Great Mind to comprehend the theory he was condemned henceforth to the Vast Majority who live on Main st. and ride in Fords.

Mr. Ginsburg offered this chaser to his cup of gall:

Anybody can master the theory if he studies long enough. The trouble is, so many die.

Take an ant crawling over a page of Shakespeare's plays. He thinks it is the boulevard link.

Raise him from the page so that his eyes focus on the lettering. Teach him English. He will appreciate Shakespeare and be grateful to you all his life, although Mr. Ginsburg did not specifically mention his gratitude.

Old Dogmas Walk the Plank

Mr. Ginsberg explained that the theory, once mastered, modified everything, particularly philosophical teachings heretofore regarded as self-evident.

Space heretofore has been considered as something that contains objects; hereafter objects will be considered as things that contain space. Time is considered to have had a beginning and therefore an end. Light is matter. When your arm has been X-rayed, matter has actually penetrated it.

After this explanation, a word as to Professor Einstein's visit. It is in behalf of the Hebrew University at Jerusalem, which he hopes to make a fountainhead of Jewish [culture] and which has already been [founded. He] will remain here several [days to lec]ture at Northwest[ern and Chicago] Universities. He [is accompanied by] his wife.

Chicago Herald and Examiner, May 3.

> This article was transcribed from a copy of an old clipping, apparently not from the so-called final edition of the *Chicago Herald and Examiner,* and its end is cropped. The microfilm copies of the newspaper available in libraries are those of the final edition; this article cannot be found in it, so I was unable to check how close my reconstruction to the original is.

"Relativity" Is Simple, Asserts Einstein, Here

"Two events, occurring simultaneously from the viewpoint of an observer standing still, may not be simultaneous from the viewpoint of a moving observer. So if you are standing still and see two flashes of lightning you may think they occur simultaneously. If you are moving between the two points, however, you will not think the two flashes were simultaneous."

There you have an easy, elementary, kindergarten explanation of the theory of relativity as explained by Professor Albert Einstein, the man who discovered it after sixteen years of study, himself. Solomon Ginsburg, his secretary, acted as interpreter.

Professor Einstein, dressed in a gray suit, and wearing a green necktie, and a big, flopping black felt hat—all the world like a country parson come to town—arrived in Chicago yesterday, accompanied by Mrs. Einstein, who frankly admits she doesn't understand what his theory is all about.

Very Simple, He Says

"Is it true that only twelve men can understand your theory?" he was asked.

"No, no," he replied, after allowing his face to indulge in a smile. "I think the majority of scientific men who have studied the theory at all can understand it."

To clear up some points of the theory that upset the scientific world, brought the universe tumbling about the ears of the uninitiated and caused more talk than Newton's apple, he obliged with some explanations.

"Rays of light bend. There is a fourth dimension. Space is not limitless. There is a beginning and end of time." Those were some of the germs to fall from his lips.

The Stars Fool You

In other words, as Mr. Ginsberg obligingly explained, according to Professor Einstein's theory, you can look over your right shoulder and see a star. Then look over your other shoulder and you'll see what appears to be another star. It isn't—it's the same star, because the ray of light is curved.

And there is such a thing as laying a tape measure around cosmos. Space is not limitless, Einstein says, and it is perfectly possible to discover by scientific means just how big the universe is.

Chicago Herald and Examiner, May 3, p. 3.

Professor Einstein Gives Three Lectures at the University

Professor Albert Einstein, of the University of Berlin, lectured at the University of Chicago on Tuesday, Wednesday, and Thursday, May 3, 4, and 5. The general subject of his lectures was "The Theory of Relativity." Professor Einstein's new theory is being widely discussed in the scientific world.

The University of Chicago Magazine, May, p. 255.

This is all I was able to dig up on Einstein's Chicago lectures. We know, however, more of his meeting with Chicago physicists, which took place not at the home of Max Epstein but of Francis Neilson, British playwright, politician, and pacifist who emigrated to the United States at the outbreak of World War I and from 1920 was editor-in-chief of the prestigious biweekly, the *Freeman.*

"When the great mathematician arrived here" writes Neilson, "we arranged a lunch for him at the house. What an unforgettable feast! He spoke no English, and I could not utter half a dozen sentences in German. But we got along well because Mrs. Einstein spoke English and Mrs. Mendelssohn could act as interpreter, when necessary. . . . But what a sense of humor he showed! I do not remember ever meeting a great man who had such a delightfully boyish appreciation of broad fun. And what a rollicking, contagious laugh; but, yet, he was extremely shy and most reserved."[52]

He introduced Einstein to two young friends at the university: Ernest Zeisler and Horace Levinson, who had been working on the mathematics of Einstein's theory.

Neilson mentions that Albert Michelson was also present. Livingston-Michelson maintains that Einstein paid a visit at Ryerson Laboratory and discussed with Michelson the value of an experiment proposed by Ludwik Silberstein; he even adds that he was shown plans in which the path of light would be projected in two directions around a rectangle (Livingston-Michelson, *Master of Light,* 290).

Even though two independent sources state that Einstein met with Michelson, the contemporary newspaper proves they are wrong. At the time Einstein was in the United States, Michelson was in Europe and, as Hendrik A. Lorentz wrote to Einstein on March 19, 1921, was going to attend the Solvay conference in Brussels starting on March 31.[53] Einstein discussed, however, Silberstein's proposal with Silberstein himself, when Silberstein volunteered to serve as translator in Princeton. Einstein was also happy to learn from him that the experiment would be performed by "the Master," Michelson.[54] For details, see p. 199. At the end of June, in his letter to Lorentz, Einstein remembered that he had met Robert A. Millikan, professor of the University of Chicago at the time but about to move to the California Institute of Technology, and that Millikan gave him an account of an important international meeting in Brussels (the Solvay conference) that Einstein could not attend because of his American trip.[55] Millikan was in Paris on an exchange program for the spring semester. It seems likely that the reporters mixed up "Millikan" and "Michelson."

Another evening the Einsteins visited a friend of Neilson's, and they arrived very early. "As the guests came filing in for an hour or so" Neilson writes, "I noticed that Einstein became more and more moody. I wondered what was wrong. Shortly afterwards, his wife came to me and said, 'Is there to be anything to eat?'" It turned out that they had thought they had been invited to dinner, and Einstein had had no time to have a normal lunch. Thanks to Neilson's intervention, the hostess did her best to have the buffet supper ready earlier than the planned 10 o'clock: "I shall never forget the look of gratitude in the professor's eyes when Elsa explained the matter to him."[56]

There is another story about Einstein's appetite, and as it is not more (or less) trustworthy than the previous one, why not serve it up here or even suppose that the storytellers are talking of the same event? When Einstein and his wife were guests of Max Shulman, the president of the Zionist Organization of Chicago, the hosts offered a second helping. Einstein, absent-minded, turned to Elsa and asked: "Have I eaten enough?"[57] L. B. Sager, the storyteller, puts this episode at about 1925. Einstein visited Chicago twice in his life, in 1921 and in 1933, so the event may have happened either year but definitely not in 1925.

On another evening Carl Beck gave a soirée for Einstein and invited musicians to meet him. After supper Einstein took part in a quintet.[58] Professor Carl Beck, director of the North Chicago Hospital and a sanatorium, had already offered his services to Einstein in 1920.[59] When he was already in the United States, Einstein apologized for not accepting his invitation to stay with him, because "my work for the Jerusalem University makes me a center of so much noise and coming and going that I would be really a most troublesome guest."[60]

"My Purpose," Said Prof. Einstein, "Is to Enlist Support of Jews of America for the Building of a Hebrew University in Palestine. Such an Institution Is Essential for Jews, So Many of Whom Are Seeking Education."

By Herman Bernstein

(Noted Correspondent, Interviewing Prof. Albert Einstein, the Famous Discoverer of the Theory of Relativity.)

Prof. Albert Einstein, who introduced a new scientific conception of space and time and of their relation to the physical world, has come to America not to expound his theory but to interest the Jews of America in the building of a Hebrew university in Palestine.

The foremost Jewish genius of our age is a modest, unassuming kindly gentleman, almost childlike in his simplicity, with a keen sense of humor.

Professor Einstein, who is a Swiss citizen and a professor of the University of Berlin, suffered from attacks directed against him by German anti-Semites when their agitation was intense. Writing to the London *Times*, he thus characterized wittily the present tendency to discriminate between men of science on nationalist grounds:

"The description of me and my circumstances in the *Times* shows an amusing feat of imagination on the part of the writer. By an application of the theory of relativity to the taste of the readers, to-day in Germany I am called a German man of science, and in England I am represented as a Swiss Jew. If I come to be regarded as a bete noire, the description will be reversed and I shall become a Swiss Jew for the Germans and a German man of science for the English."

Help Upbuild Palestine

I interviewed Professor Einstein for the *New York American* on the purpose of his mission in America and the needs of the University of Jerusalem. Answering my questions, Professor Einstein said:

"The purpose of my visit to America at this time is to enlist both the moral and material support of the Jews of America for the building of a Hebrew university in Palestine.

"Such a university would assist enormously in the upbuilding of Palestine, for it would become a spiritual center of Jewish education and culture for the Jews in the Holy Land. Later this university would also attract Jewish students in other lands, who would come to study in Palestine.

"But, above all, I consider it most important that there exists this great common Jewish cultural enterprise, which is of the utmost significance for all Jews.

"I have arrived at the decision that such an institution of learning in Palestine is essential through my contact with Jewish students in various parts of Europe.

"I have met the Jewish students of Austria, Hungary and of Russia, who are clamoring to complete their education in European universities. I have seen the difficulties and hardships which these young men and women are experiencing. I have observed the discriminations, which have become intensified since the war.

Fans Flames of Hatred

"Poverty, the economic collapse in Central European countries, has fanned the flames of national hatreds, and the Jewish students have suffered in Germany, Poland and especially in Hungary. A large number of Jewish students are barred from the universities on religious grounds.

"The situation has been aggravated also by the fact that a number of universities in Eastern Europe have been closed altogether.

"Many of the students have come to me for help. I have seen their help-lessness, their hardships and their needs. And in this way I came to the realization that a Hebrew University in Palestine is an absolute necessity for the Jewish people.

"The traditional respect for knowledge which we Jews have maintained intact for many centuries of severe persecution makes us feel all the more keenly the present discrimination against so many talented sons and daughters of the Jewish people who are knocking in vain at the doors of the universities of Eastern and Central Europe.

"And those who have gained access to the spheres of free research had to do so by undergoing a process of assimilation which has crippled the free and natural development of the spiritual character of our people and deprived them of their cultural leaders.

Preserve Hebrew Arts

"It seems to me it is also the duty of the Jewish people to preserve, through the university we are planning to establish in Palestine, the neglected branches of Hebrew literature, language, archaeology and history.

"Distinguished Jewish scholars in all branches of learning are awaiting to go to Jerusalem where they will lay the new foundation of a flourishing spiritual life and will promote the intellectual and economic development of Palestine.

"The Hebrew university in Palestine would become a new 'holy place' to our people.

"Despite the crude realism and materialism of our times, there is a glimmer of a nobler conception of human aspirations. The American people exemplified this by the part they played in recent years in the affairs of the world.

"The Jews of America are at this time the most fortunate portion of the Jewish people. Europe is sick and suffering, and the Jews of Europe are experiencing greater sufferings, discriminations and persecutions than ever before.

"So I have come here with feelings of hope that my spiritual aims with regard to the university in Jerusalem will find a sympathetic response in America, and will be realized through the support of American Jewry.

"A group of physicians has been formed for the purpose of raising funds for the medical faculty of the university. This organization alone has undertaken to raise a fund of a million dollars in the course of two years.

Build without Delay

"A general committee is also being organized, including a number of prominent academicians and financiers. With the help of these people I hope to secure the required moral and material support to enable us to build the university without delay.

"Not only Zionists are realizing the importance of this university, but a number of distinguished American Jews not identified with the Zionist movement have expressed their interest and sympathy for this cultural enterprise and have promised to assist me in making it a success.

"I am not an organizer myself. The university will be organized by specialists. I shall be glad to work with them and help them in every way possible. I am at the disposal of the university, and am prepared to participate in the scientific department.

"Not being a prophet, I cannot foretell whether my mission in this country will be successful. All I can do is to do what I can. But the sympathy and interest for this work I have found in this country both among the Jews and the Gentiles is most encouraging.

"Various considerations have led to the selection of the following university institutes for the initial scheme:

"1.—A department of Jewish and Oriental studies—philology, literature, history, law, archaeology, religion and philosophy, mainly Jewish but including also Arabic and Semitics in general. This department is to be a university school for scientific studies, able to offer training to both graduates and postgraduates, and empowered to confer degrees.

A Research Institute

"2.—A research institute for the Hebrew language, the object of which will be to guide and assist its modern development by the study of its vast treasure house of literature.

"On the scientific side it was decided to begin with research institutes, as suggested in 1913 by Dr. Weizmann and the university committee, in which the chief scientific adviser was the late Prof. Paul Ehrlich, and not with teaching faculties. The initial plan comprises the institutes of physics, chemistry and microbiology."

Einstein with group below the telescope of Yerkes Observatory, Williams Bay, Wisconsin. Yerkes Observatory, courtesy AIP Emilio Segrè Visual Archives.

"In the advancement of science Jews have always taken a noble part, but the fruits of their labors have not been reaped by Jewry.

"Is it conceivable that, in addition to the tragedy of Jewish science without a home, there could exist a Jewish national home without science? The traditional pride of the Jewish people in their learned men would never suffer such humiliation.

"And there is no doubt that to those non-Jewish idealists and believers in spiritual values who have supported Zionism a Jewish Palestine means, perhaps mainly, a real renaissance of that Jewish genius of which they have seen so many examples scattered in many lands."

Chicago Herald and Examiner, May 8, part 2, p. 3.

Friday, May 6

On this day Einstein paid a visit to Yerkes Observatory. The director, Edwin B. Frost, sent a photo to Max Epstein on May 20 and added that "all connected with the University of Chicago as well as many people outside of our institution, have greatly enjoyed the opportunity which you gave us of hearing and meeting Dr. Einstein, who left a most agreeable impression of his personal character as well as of his great intellectual power."[61]

13

Princeton Hears Einstein Explain

✦

Friday, April 8

Professor to Deliver Five Lectures in Princeton University

Special Telegram to *Public Ledger*

PRINCETON, N.J., April 8.—Prof. Albert Einstein, of the University of Berlin, will give a series of five lectures at Princeton University, beginning on Monday, May 9. The professor will be the guest of the university during his stay here. In all probability scientists of other universities will be asked to attend the lectures.

The lectures, which will be the fullest account of the theory of relativity ever presented in America, will be delivered in German, and will be interpreted for the audience.

The Public Ledger (Philadelphia), April 9, p. 3.

> President John Grier Hibben asked Einstein through Paul Warburg whether Einstein would think it useful to take the lectures in Princeton in shorthand.[62] The problem was aggravated because "Dr. Einstein speaks practically no English, although some French," as Gano Dunn wrote to Oswald Veblen, professor of mathematics at Princeton University. The organizers hoped that Dunn would know where to find a German stenographer. "I don't know of any expert German stenographers," Dunn answered, "but I have heard several of Dr. Einstein's lectures . . . and I have also talked to him a good deal and recognize that his diction and enunciation are unusually clear and simple, rendering a stenographer's work much easier than in the case of most German speaking individuals."[63]

President Hibben was of the opinion that the published lectures would sell well and offered Einstein a 15 percent royalty.[64]

Friday, April 22

Lecture by Professor Einstein

... Princeton University has arranged five lectures on the theory of relativity on the afternoons from May 9 to 13, the subject of these lectures, which will be delivered in German, are first and second "Generalities on the theory of relativity," (without going deeply into the mathematical symbolism); third "Special theory of relativity," fourth "General theory of relativity and gravitation," fifth "Cosmological speculations." Scientific men are invited to the lectures. Admission will be by ticket, application for which should be forwarded to Professor H. D. Thompson, Princeton, N.J.

Science, April 22 (53 [1921]), p. 382.

Saturday, April 30

Einstein Invites 600 to Cosmological Party

PRINCETON, N.J., April 30.—America tonight was offered a chance to expand the sacred circle of those who understand the Einstein theory of relativity.

Invitations were sent out tonight to more than 600 college and university presidents to attend a series of lectures, beginning May 9, when Dr. Albert Einstein will explain his theory in person.

Five lectures will be given, beginning with an introduction to the theory and closing with "Cosmological Speculations."

The New York Call, May 1, p. 6.

Monday, May 9

Albert Einstein Speaks on Theory Here To-day

Address in McCosh 50 Will Follow the Official Welcome Ceremony in Alexander

Professor Albert Einstein of the University of Berlin will deliver the first of his series of lectures on the "Theory of Relativity" this afternoon in 50 Mc-Cosh. There will be an official welcome for the physicist at 4 in Alexander Hall.

Immediately following this formal ceremony Professor Einstein will deliver his opening address. There will be no interpreter present, but after the lecture Professor Adams of the University Faculty will give a short resumé of what has been said.

Faculty Will Attend

At the formal welcome this afternoon, there will be present a gathering of the foremost scientists of the country. Invitations have been extended to over 600 presidents of colleges and universities, physicists, engineers, and astronomers.

The University Faculty will attend this formal reception in full academic costume. During the five days that the Berlin professor discusses his theory in Princeton, visitors who intend to stay over will be accommodated at the Seminary, and will be the guests of the University.

Lectures Are in German

The other lectures of Dr. Einstein's series will be given in McCosh 50 at 4:15 Princeton time. The lecture to-day will be an introductory explanation of the relativity theory and will be given under the subject "Generalities on the Theory of Relativity." After a continuation of this general theme to-morrow, Professor Einstein will enter more deeply into the subject.

The topics Wednesday and Thursday are: "The Special Theory of Relativity" and "The General Theory of Relativity and Gravitation." The series will end Friday afternoon with an explanation of "Cosmological Speculations." Each of Professor Einstein's lectures which will be given in German, will be followed by a resumé in English.

The Daily Princetonian, May 9.

Princeton is honored this week

by visit of Professor Albert Einstein, who is delivering the course of lectures we have already announced, on "The Theory of Relativity." A number of distinguished scientists are also in Princeton to hear the lectures. Professor Einstein was officially welcomed by the University with ceremonies in Alexander Hall on Monday afternoon, at which the honorary degree of Doctor of Science was conferred upon him. Preceding the conferring of the degree, President Hibben, speaking in German, welcomed Professor Einstein to Princeton. The following is a translation of President Hibben's greeting:

"Highly Honored Professor and Colleague:

"It is a particular honor and a highly esteemed privilege to bid you a hearty welcome to Princeton.

"Naturally we have heard much about the famous Theory of Relativity, and we have also read somewhat. But have we understood? That is a question. Today we have the fortunate occasion of listening to your own explanation of this theory, and we are confident that through your deft leadership we shall find a ray of light cast upon this dark pathway.

"In your writing you have with daring thought speculated upon the possibility of a finite and yet unlimited universe. Whether this universe is finite or infinite it is not for me to say. Certainly, however, there is a world which has no limits whatever. This is the world of the spirit, to which you belong by unquestioned right. This world the great German philosopher Hegel has described as follows:

"'Man cannot form a sufficiently lofty idea of the greatness and power of the spirit. The hidden essence of the universe has no power in itself which can offer any resistance to the desire for knowledge. It must unblock itself to the seeker, lay its riches before him, disclose its depths to his eyes, and lay itself open for his enjoyment.'"

In presenting Professor Einstein for the honorary degree, Dean West said:

"Albert Einstein, professor and member of the Berlin Academy of Sciences, now professor in the University of Leyden, the foremost scientific theorist of our age. We can indicate only slightly his far-reaching achievements. He has advanced beyond all others by extending the quantum theory to the relation of electricity to matter, deepening the foundations

A cordial invitation is extended to you, and to any of your associates who may be interested, to attend the series of five lectures on The Theory of Relativity which will be delivered by Professor Albert Einstein in Princeton University on the afternoons of May ninth to May thirteenth inclusive.

JOHN GRIER HIBBEN,
President of Princeton University.

You and such of your associates as can be present are invited to appear in academic costume at Nassau Hall promptly at ten minutes past three o'clock, standard time, Monday afternoon, May ninth, in order to join in the academic welcome to be given by Princeton University to Professor Einstein before the delivery of his first lecture.

JOHN GRIER HIBBEN,
President of Princeton University.

(The hours given are standard time.)

The subjects of Professor Einstein's lectures, which will be delivered in German, are:

Monday, May 9, 4 p. m., Tuesday, May 10, 3.15 p. m. Generalities on the Theory of Relativity (without using mathematical symbols.)

Wednesday, May 11, 3.15 p. m. Special Theory of Relativity.

Thursday, May 12, 3.15 p. m. General Theory of Relativity and Gravitation.

Friday, May 13, 3.15 p. m. Cosmological Speculations.

The last three lectures are addressed to the scientific public. Admission will be by ticket, application for which should be forwarded as soon as possible, using the form enclosed.

Invitation packet for Einstein's Princeton lectures.
Courtesy Stanford University, Department of Special Collections, Manuscript Division.

(To be sent to Professor H. D. Thompson, Princeton, N. J.)

I wish a ticket, (or tickets,) for Professor Einstein's lectures on the dates underscored below:

	Monday	Tuesday	Wednesday	Thursday	Friday
May	9	10	11	12	13

Name_____

Address_____

- -

(To be sent to Professor H. D. Thompson, Princeton, N. J.)

A place in the Academic Procession will be reserved for you, if you return this card.

_____

I shall be present at Nassau Hall in academic costume, (cap, gown, and hood), at ten minutes past three o'clock, Monday afternoon, May ninth, 1921.

Signed_____

Address_____

Degree or Academic Institution_____

· ·

A limited number of men can procure accommodations in one of the dormitories in Princeton at two dollars a bed a night. Application for these rooms should be made as early as possible, and addressed to Professor H. D. Thompson, Princeton, N. J.

_____

Please reserve for me a bed in one of the Seminary dormitories at two dollars a night for the nights underscored below:

	Monday	Tuesday	Wednesday	Thursday	Friday
May	9	10	11	12	13

I should like to have a room with Mr._____

Name_____

Address_____

of thermodynamics, opening the way for a more intimate knowledge of atomic structure and bringing to its fullness the theory of special relativity. These labors alone reveal the work of a giant. *Ex pede Herculem.*

"But his highest intellectual fame rests on his general theory of relativity, a new, profound and daring conception of space and time and of their relation to the physical universe, affecting our theories of light and gravitation and joining the two mysteries of inertia and gravitation in one yet more elemental conception.

"Strict experimental proofs in the field of planetary motion, and in regard to his startling prediction that light when passing the sun is bent in its course, have already convinced the scientific world that his conceptions open a new era for progress in the entire range of the physical sciences. And not the least lesson he has taught is that the older sciences, by reason of their deeper rootage, are capable of farther thrusts than less developed sciences can usually make into the deeper recesses of Nature. In his structural theory of our ever-old, ever-new universe his name stands latest in that illustrious series wherein the other moderns are Clerk Maxwell, Isaac Newton and Galileo, and the earliest name is Pythagoras. Perhaps as a new Pythagorean his fine sensibilities in music have helped him to hear more perfectly the music of the spheres.

"And especially would we do him honor for the moral fidelity whereby, amid distressing perils, he refused to join with others in condoning the invasion of Belgium.

"So today for his genius and integrity we, who inadequately measure his power, salute the new Columbus of science,

'Voyaging through strange seas of thought alone.'"

Princeton Alumni Weekly, May 11, pp. 713–14.

In *Presentations for Honorary Degrees in Princeton University, 1905–1921*, Einstein's titles in West's introduction are corrected as follows: "Albert Einstein, Director of the Kaiser Wilhelm Physics Institute, member of the Royal Prussian Academy of Sciences, Professor of Physics in the University of Leyden."

West, "63 inches around the vest," as the undergraduates characterized him, a wit and a satirist, took special pleasure in writing honorary degree citations and in presenting them with his impressive voice.

Einstein Receives Princeton Degree

Relativity Exponent Is Made Doctor of Science in Presence of Distinguished Savants—"Successor of Newton"—Dean West Hails Guest as "Intellectual Giant"—Hibben Welcomes Him in German

Special to *The New York Times.*

PRINCETON, N.J., May 9.—Professor Albert Einstein, propounder of the theory of relativity, was honored today by Princeton University, which conferred on him the Degree of Doctor of Science in the presence of distinguished scientists and educators from many colleges and scientific institutions in the East, including six college Presidents. The academic procession, bestowed only on very distinguished visitors, was the largest and most imposing in many years.

Dean Andrew West, in presenting Professor Einstein for the degree, referred to him as an intellectual giant, the legitimate successor of Newton and Galileo, and "the foremost scientific theorist of our age." Professor Einstein was visibly moved by the tribute to his work, and to himself as one who had refused to join in condoning the invasion of Belgium. In a brief speech he expressed his appreciation of the honor paid him.

He arrived at Princeton at 4 o'clock and was met by President John Grier Hibben. After a short visit to Nassau Hall, where members of the faculty awaited him, he donned his robe. Escorted by President Hibben, he walked at the head of the academic procession to Alexander Hall. The sombre gowns of the faculty, some relieved by the bright colors denoting their degrees, made a picturesque and impressive picture as they passed under the green trees that line the campus walks.

Among the visitors who took part in the procession were President W. A. Neilson of Smith College, President R. H. Crossfield of Transylvania University, President Charles W. Flint of Cornell, President Parke R. Kolbe of Akron University, President H. J. Pearce of Brenau College, President Cyrus Adler of Dropsie College, Professor C. H. Currier of Brown University, Arthur L. Day of the Geophysical Laboratory, Washington; Professor Arthur W. Goodspeed of the University of Pennsylvania, Professor C. M. Gordon, of Lafayette College, Professor J. M. Young of Dartmouth College, Professor I. J. Schwatt of the University of Pennsylvania, Ludwig Silverstein of Rochester, Headmaster M. A. Abbott of Lawrenceville, Professor L. G. Poaler of Columbia, Dr. A. Pell of Bryn Mawr, Dr. L. S. McDowell of Wellesley, Dr.

Einstein in procession on the occasion of his receiving an honorary degree at Princeton University. Central Zionist Archives, Jerusalem, courtesy AIP Emilio Segrè Visual Archives.

J. L. Luck of the University of Virginia, Dr. T. D. Cope of the University of Pennsylvania.

[Here follows a long quotation from Hibben's and West's speeches, already cited above, although here they are attributed to Hibben only.]

President Hibben then pronounced the Latin words conferring on Prof. Einstein the degree, and Colonel William N. Libbey, who acted as marshal of the procession, placed the academic hood around his neck.

The New York Times, May 10, p. 17.

On May 10, the *Public Ledger* of Philadelphia ran a notice of the event, and the *Daily News* even added excerpts from Adams's resume, which appears in its entirety on pp. 170–171, 174–177.

Degree Is Awarded Albert Einstein at Academic Welcome

Noted Scientists from Many Institutions Gather to Honor Promulgator of Theory of Relativity of Matter—Shows Time and Space Are Not Absolute But Relative—Continues Series of Lectures in German This Afternoon at 4:15 in McCosh 50 on Generalities of Theory

Yesterday in Alexander Hall Princeton gave its welcome to Dr. Albert Einstein, the man who stands as the greatest scientist of the age, ranking with Pythagoras, Galileo, and Newton. The reception took place at 4 at which time Dean Andrew Fleming West of the Graduate College presented on behalf of the University the degree of Doctor of Science to Professor Einstein. Following this, Dr. Einstein, now professor at Leyden, spoke in McCosh 50 in German on "Generalities on the Theories of Relativity." Today he will speak at 4:15 in McCosh 50, continuing with the same subject.

Einstein Awarded Degree

President Hibben opened the reception with an address of salutation to the distinguished guest, paying a high tribute in which he said: "In your writing you have with intrepid thought speculated upon the possibility of a finite and yet unlimited universe. Certainly, however, there is a world which has no limits whatever. This is the world of spirit to which you belong by unquestioned right."

In conferring the honorary degree, Dean West spoke of the courage and far-reaching exploits of this Columbus of science which reveal the work of a giant "voyaging through strange seas of thought alone." Dr. Einstein in his now famous theory of relativity has expressed a profound and daring conception of space and time and opened a new era for progress in all the physical sciences. In refusing to condone the invasion of Belgium, he displayed the same sincerity of conviction which has characterized his theories.

The Lecture

At the conclusion of the reception in Alexander Hall, Colonel Libbey led the procession of several hundred of the foremost scholars of the day, who had been invited to Princeton, to McCosh 50. Here Dr. Einstein delivered his lecture in his native tongue, President Hibben presiding. The hall was nearly full, and it was difficult for those in the rear to hear the speaker who spoke in rather a low voice. After the talk, Prof. E. P. Adams of the University Faculty, gave a resumé in English of the scientist's remarks.

A study of the question, "Is there any state of motion that has the unique property of being completely at rest?" would constitute a study of the law that Einstein promulgates, and which in his interpretation, answers no. Sir Isaac Newton and Galileo believed that time and space are absolute. Professor Einstein has conclusive proof that they are only relative. The distance between two points in space, for example, is not absolute, but depends upon the position and the motion of the observer.

Study of Motion

Or putting the statement in another way, if someone saw a wagon moving down a street, could he say that the wagon was in motion and the street at rest, or the wagon at rest and the street in motion? The classic physicist would agree that the street was the only object at rest, but Einstein says that the matter can be considered either way. For, according to his theory of relativity, the only system absolutely at rest is that one ordinarily spoken of as ether.

In the study of motion there are two laws which govern the action of all natural phenomena: the principle of relativity, and the constancy of the velocity of light. The first law states that all inertial systems are equivalent in expressing laws by which natural phenomena are measured and governed. This is assuming that motion in no way affects the size of an object, but under Einstein's theory, the length of a rod moving *in the direction of its length* gets shorter. If it moves with the velocity of light, its length is zero.

Time and Space Relative

The velocity of light is, therefore, a limiting velocity above which an object can not go and still retain its size. This velocity is absolute. Hence if an observer stands at an equal distance from two flashes, he does not necessarily see them at the same time unless he is standing still. They do not appear simultaneous if he is moving.

This one fallacy which assumes that motion in no way affects the size of an object was not perceived by the scientists of the old school; in measuring time and length of one system in respect to another, these physicists would consider both time and space absolute. But Einstein considers both time and space relative on the basis that both are dependent upon the velocity with which light is traveling.

The Daily Princetonian, May 10, pp. 1–2.

> I hope that you were surprised to read that Einstein was professor not only in Berlin but also in Leyden. The year before, he was appointed special professor there. Earlier, his friend Paul Ehrenfest and his Dutch mentor Hendrik A. Lorentz offered him a permanent position as a safe haven after the news about the scandalous anti-relativistic and anti-Semitic events had reached them. Einstein did not accept the permanent position and remained faithful to his Berlin colleagues, but as special professor he visited Leyden time and again.[65]

Princeton Honors Fuss Dr. Einstein

Theory of Relativity Easier for German Scientist Than Getting a Degree—Grins at His Mistakes

> Albert Einstein has come around
> To tell Princeton what he's found
> That Isaac Newton had unsound
> Ideas of gravitation.

It is easier for Dr. Einstein, the Christopher Columbus of Space, to determine relative positions a few million light years out in the universe than to get the right relative position when receiving a degree from a college president.

This hypothesis is built upon the maneuvers of Dr. Albert Einstein, the German scientist and promulgator of the Theory of Relativity, yesterday, when he received the honorary degree of Doctor of Science at Princeton University.

When Dr. Einstein was supposed to sit down he stood, he sat when he should have stood and finally when the smiling, confused scientist who does not understand English, steered about by little tugs at his sleeve, was to receive his hood of a doctor from President John Grier Hibben he turned his back in the confusion upon the president.

There were six college presidents and representatives of about twenty colleges in the academic procession to Alexander Hall, where the degree was conferred. The tugs were hidden behind Dr. Einstein's gown, and the signals to sit down and stand were whispered so low that they barely reached the front of the platform.

"In your writing you have with intrepid thought speculated upon the possibility of a finite and yet limited universe," said President Hibben, speaking in German. "Whether this universe is finite or infinite it is not for me to say. Certainly there is a world of the spirit, to which you belong by unquestioned right."

In McCosh Hall, where the first of five lectures was delivered, there were about 400 scientific men and women crowded for the first talk.

The scientist picked up a piece of chalk before he began to talk. Part of the theory he explained on the blackboard. The rest he explained out in space, with the chalk drawing imaginary lines.

The long hair that ended in tight curls and the chalk balanced between his fingers like the baton gave him the appearance of an orchestra leader when not at the blackboard.

He wore a green knitted tie with a thin red stripe in it, perfectly creased pin-stripe trousers and a formal black coat. Some crumbs of thought offered by Dr. Einstein that will require considerable digestion to become street gossip are:

"If we have a clock in one system which we will say it is a system at rest and then refer that clock to a system that is moving with uniform velocity with respect to the system at rest—the clock is found as a result of the theory to go slower in the second system than it did in the first.

"If the length of a rod is moving in the direction of its length, it becomes shorter in the direction of its length. If it is moving transversely then there is no change."

In other words, it was explained by Professor Edwin P. Adams, of the physics department, our world when viewed from out in space by a body properly related in its motion, it would appear as an ellipsoid or a world that looked as if someone had sat on it.

Distances are relative. It all depends upon how you look at a thing, explain professors who heard the first lecture.

The Evening Bulletin—Philadelphia, May 10, p. 1.

Princeton Hears Einstein Explain

Scientist Delivers First of Lecture Series—Declares Laws of Mechanics Do Not Satisfy Theory of Relativity

(Special Dispatch to the *Evening Post*)

PRINCETON, N.J., May 10.—Princeton University paid tribute to Dr. Albert Einstein in a formal academic welcome in Alexander Hall yesterday afternoon, at which time the degree of Sc.D., *honoris causa,* was conferred upon him. A group of distinguished American scholars were present at the exercises and also attended the first of Dr. Einstein's series of five lectures.

Dr. Einstein and his wife arrived in Princeton early in the afternoon from Chicago, and at 4 o'clock the academic procession, headed by the university marshal, Col. William Libbey, moved from Nassau Hall to Alexander Hall. Dr. John Grier Hibben, president of the university, walked beside Dr. Einstein, and following them were the visitors from other university faculties, with the Princeton professors and graduate students bringing up the rear.

After a brief introduction Dr. Hibben turned to Dr. Einstein and said in German: "It is a particular honor and a highly esteemed privilege to bid you a hearty welcome to Princeton.

"Naturally we have heard much about the famous theory of relativity, and we have also read somewhat. But have we understood? To-day we have the fortunate occasion of listening to your own explanation of the theory."

Among the American scholars who were present at the reception to Dr. Einstein and who later attended the lecture were the presidents of six colleges and members of the faculties of a score or more other institutions. Those on hand included: W. A. Neilson, president of Smith College; R. Crossfield, president of Transylvania University; Charles Flint, president of Cornell College; Parke R. Kolbe, president of Akron University; H. J. Pearce, president of Brenau College; Cyrus Adler President of Dropsie College. Dr. L. G. Poaler of Columbia University; Prof. H. C. Currier of Brown University; Prof. Arthur W. Goodspeed of University of Pennsylvania; Prof. C. M. Gordon of Lafayette; Prof. J. W. Young of Dartmouth; Prof. I. J. Schwatt of University of Pennsylvania; Mather Abbott, headmaster of Lawrenceville; Dr. A. Pelly of Bryn Mawr; Dr. L. S. McDowell of Wellesley College; Dr. J. L. Luck of the University of Virginia; Dr. T. D. Cope of the University of Pennsylvania; Ludwik Silverstein of Rochester, New York, and Arthur Day of the Geophysical Laboratory, Washington.

After the ceremonies in Alexander Hall Dr. Einstein discussed "Generalities on the Theory of Relativity," on which topic he will also talk this afternoon. At the close of the address Prof. Edwin Plimpton Adams of the physics department of the university made a half-hour résumé. As summarized by Prof. Adams, Dr. Einstein said in substance:

Dr. Einstein's Explanation

"What we mean by motion, relative motion in a general sense, is perfectly plain to every one. If we think of a wagon moving along a street, we all know that it is perfectly possible to speak of the wagon at rest and the street in motion, just as well as it is to speak of the wagon in motion and the street at rest. That, however, is a very special part of the ideas involved in the principle of relativity.

"The question is this: 'Is there any unique state of motion—we can speak of it in that general way—which corresponds to absolute rest?' In other words, is there a certain thing which we can speak of and say that it is absolutely at rest? And the answer to that question involves the whole theory of relativity.

"Now, to get ahead, it is necessary to speak just slightly of the historical development of the thing. In the ordinary treatment of mechanics, particularly in the laws of motion which follow from the principles of Galileo and Newton, we know perfectly well that we can speak of all systems if they are moving with respect to each other in uniform and rectilinear motion as equivalent systems. In other words, the laws of mechanics are just the same if we refer them to a system which is moving with uniform and rectilinear velocity with respect to another system. If, however, we refer to a system which is moving with a rotational motion with respect to another system, then it is well known that the laws of mechanics are not the same as they are when referred to a system at rest.

"What we mean by an inertial system, then, in the special theory of relativity is a system which moves with respect to another system with a uniform velocity in a straight line.

"The principle of relativity states that all inertial systems—that is, all systems which move with uniform and rectilinear velocity with respect to each other—the [are] equivalent in expressing the laws of natural phenomena.

"The second law that we state is the law of constancy of the velocity of light. Now these two laws are experimental laws; the experiments which prove these are indirect, but they are hypotheses which we make, and then we try to find out how nearly all natural phenomena fall in with them.

Some Striking Contradictions

"Now there are first of all certain very—there are to be—very striking contradictions between these two principles and the principles derived from the assumption that there is one system which has the unique properties of being absolutely at rest. The simplest way to think of it is to suppose that you start a ray of light referred to a system which we can say is absolutely at rest. Say that it goes a distance C in one second. Now, if it is referred to a system which is moving with the velocity V then, referred to the second system it would travel with a velocity of C plus or minus V, depending upon whether the ray is moving in the direction of V or opposite.

"That, however, which is a consequence of assuming that there is a system which has the unique properties of being at rest, is directly in contradiction to the principle we stated of the constancy of the velocity of light.

"Now, this is capable of experimental verification, and it was tested by the famous experiments of Michelson, who measured accurately the velocity of light when travelling in the direction of the earth's motion and also traveling at right angles to the direction of the earth's motion. According to our ordinary ideas, we would expect to find a difference in velocity of light in the two cases. Michelson found absolutely no difference. And therefore there is a contradiction between the two principles that we stated, the principle of relativity and the principle of the constancy of the velocity of light, with the assumption that there is one unique system which we can speak of as a system at rest.

"The only thing that we are left with, then, is to assume that the assumptions that we have made in formulating our ideas of motion are fundamentally wrong. And the particular features of our fundamental ideas which we have got to change are, first of all, the definition of what we mean by two simultaneous events. If we have two events which take place at different places, what exactly do we mean when we say that these two events are simultaneous? In order to test whether they are simultaneous or not we have to make use of the principle of the propagation of light.

"Now, the definition which we shall introduce in order to define simultaneous events depends upon our assumption of the constancy of the velocity of light. Suppose that we have two events taking place at different places, and if we consider an observer placed just half way between the two events, we then say that the two events are simultaneous if that observer notes that those two events are simultaneous for him, and that definition is to be the definition of simultaneity of events.

"Now there is another assumption which we have made in our ordinary ideas, and that has to do with measured lengths. We ordinarily suppose that the length of an object that we measure is wholly independent of its state of motion. That, however, is purely an assumption, and whether it is so or not will have to be tested by the consequences of the special theory of relativity.

"Everything comes to this. We refer all physical phenomena to a system of coördinates and to a way of measuring time. Suppose that we have our physical phenomena referred to some particular system of coördinates. The special theory of relativity states that physical phenomena referred to any other system of coördinates which are moving relatively to our first system with uniform and rectilinear motion must be expressed by exactly the same laws. And that statement enables us to find out what measured lengths in one system will be when measured with respect to our second system. It also enables us to find out how measured time in one system will come out when measured in respect to another system.

"The thing that we have to do is to express mathematically the relation between the coördinates and time measured in one system which we can say is at rest, and another system in motion. It was Lorentz who first found how all measurement of lengths and time referred to one system must be expressed when we wish to refer them to another system.

"And that, then, is the principle of relativity which states that the laws of all physical phenomena must be of the same form when referred to two different systems which are moving relatively to each other with uniform rectilinear velocity.

Consequences of Theory

"Now, the consequences of this transformation from one system to another in brief are these: First of all, it is found that the length of a rod, if it is moving in the direction of its length, becomes shorter in the direction of its motion. If it is moving transversely to its length, then there is no change in its direction. In particular, a sphere in motion becomes flattened out in the direction of motion. In other words, it becomes an ellipsoid.

"Another consequence is that time—if we have a clock in one system at rest and then refer that same clock to a system which is moving with uniform velocity with respect to what we speak of as our system at rest—that clock is found as a result of the theory to go slower in the second system than it did in the first system.

"Another important consequence of the application of this principle to physical laws is this: That the velocity of light is a limiting velocity, above which we cannot go. In particular, if we have two systems, one system, we will say, is moving with a velocity which is very nearly but not quite the velocity of light, and then consider a third system which, with respect to the second system is also moving with a velocity which is very nearly but not quite the velocity of light, our ordinary notion would be that the first system would be moving in respect to the third system at a velocity which is very much greater than the velocity of light.

"The consequence of the principle of relativity is that when velocities are added in this way we will always find a velocity which is less than the velocity of light.

"Now, we have this consequence finally, that the laws of all physical phenomena must satisfy the conditions as given by the principle of relativity, that the statement of all of the laws of physical phenomena must be the same whether they are referred to a system which we can speak of as at rest or referred to a system which is moving with uniform rectilinear velocity. The laws of electro-magnetism and electro-dynamics in general satisfy this condition. The laws of mechanics, however, in the form in which we are used to them, do not satisfy this criterion."

New York Evening Post, May 10, p. 7.

Professor Einstein's Princeton lectures on "The Theory of Relativity"

will be published in book form by the Princeton University Press. Professor Einstein has agreed not to authorize the publication of his lectures in the United States by anyone else, so that this volume will contain not only the latest exposition of the Theory of Relativity, but it will be the only one authorized by the famous scientist during his visit to this country. A German stenographer is taking notes of the lectures as they are delivered. The plan of procedure is to have her write her notes out in German and then Professor Edwin P. Adams of the Department of Physics will go over them and check up those scientific portions which may have caused trouble. After Professor Adams has completed this part of the work, the lectures will be submitted to Professor Einstein for revision and final approval. When he has returned them, they will be translated into English and published. The Press feels that it is to be congratulated on having been able to make arrangements for this book, which will be one of the most important scientific works of recent years.

Princeton Alumni Weekly, May 11, pp. 713–14.

A Memorandum of Agreement was settled on May 9 between Einstein, "Professor in the University of Leyden, Germany," (!) and Princeton University Press and signed by him and Paul G. Tomlinson, Manager of the Press, on the publication of a work entitled "Lectures delivered at Princeton University, May 9–13, 1921."[66]

The shorthand notes for the first lecture and first half of the second lecture, with no mathematical formulas, were published in volume 7 of the *Collected Papers.* Eventually the lectures were published in 1923 under the title *The Meaning of Relativity: Four Lectures Delivered at Princeton University.* The publication includes the five lectures he actually delivered. We don't know how much use he made of the stenographer's notes in the final version.

Einstein Feels at Home

Prof. Albert Einstein received the degree of Doctor of Science from Princeton University last Monday. He has been addressing audiences of scientists at Columbia and City College and at Chicago. He is feeling as much at home on these occasions as he is very much at sea in the Zionist embroglio. What a pity that he permitted himself to be used as the tail to the Zionist kite, instead of coming to this country on the invitation of great institutions of learning, as would have become the scientist that he is! The European Zionist Commission, in taking Einstein unto itself at the last moment, demonstrated at the outset that its leaders had not the faintest conception of the American spirit. Of course, it was far from their thought that the silent professor could become the hero of the press and people, while the loud orators would be left to the mercy of the Yiddish press. Einstein will not talk on Zionism; that is left to Weizmann, Ushishkin and Levin. Thank heavens, that while, relatively speaking, the latter are spending a modest hundred millions in hot air, Einstein is given the leisure to talk Relativity to those who hope to understand him.

The American Hebrew, May 13, p. 773.

Tuesday, May 10

Einstein Says Theory Is Liable to Collapse

Permanency Depends upon Validity of Conclusion Regarding Vibration from Sun—Bent of Spectral Lines Will Prove or Disprove—Unless They Lean Slightly Toward Red of the Spectrum, Relativity Is a Fallacy

Dr. Albert Einstein admitted yesterday that his theory of relativity would completely collapse if one of his conclusions can be disproved. The conclusion referred to was the one regarding the vibration coming from the Sun through the chemical atoms filling space. "These vibrations," said Dr. Einstein, "would have their spectral lines bent slightly toward the red of the spectrum. If this is untrue, the whole theory is a fallacy." The talk to-day will be held at 4:15 in McCosh 50 on "The Special Theory of Relativity."

Dr. Einstein Broadminded

It was one of the startling admissions of yesterday's lecture by Dr. Einstein, who has been extremely broadminded throughout all his conclusions and frankly acknowledges the possibility of his error, though he himself is convinced of its truth. The talk took place in McCosh 50 at 4:15 and was followed, as before, by a resumé by Prof. E. P. Adams of the Faculty. In the first lecture Einstein took up the special theory of relativity which depends upon the hypothesis that light waves travel with the same velocity whenever referred to an inertial system—a system which moves with a velocity which is uniform and rectilinear.

But the striking thing here is that this apparently applies only to systems with a particular type of motion. How are we to deal with systems having a perfectly arbitrary type of motion which does not come under that category? The *general* theory of relativity attempts to explain this, using the principle of inertia. In respect to any one of our *inertial* systems a simple mass point, if it is far enough away from another body, moves with uniform velocity in a straight line. But this principle of inertia does not hold when we refer the mass point to a system moving in *arbitrary* motion, such as a motion of rotation. In that case, instead of the motion of the particle that we considered first being in a straight line, it may in certain cases be a very complicated curve.

Illustrated by Disk

Now the whole foundation of scientific physics must be changed if this system of arbitrary motion is going to be equivalent to the inertial systems. This is the principle of equivalence which is the foundation of the general principle of relativity, and this negates the rules laid down by Euclidian geometry. This can be easily seen by the illustration of a disk in uniform rotation in a plane. Under Euclidian geometry with a system of co-ordinates at rest, the circumference and the diameter of a circle could be measured by extremely small, exactly identical measuring rods placed end to end along the diameter and along the circumference. The ratio of the number of these measuring rods in the circumference to the number in the diameter will be *pi* according to Euclidian geometry. If, however, the disk is in motion, according to the special theory of relativity explained in the first lecture, the ratio is greater than *pi* for the reason that some of the measuring rods which are in the circumference will be moving along the direction of their length, and such rods, according to the theory, are always shortened.

Thus the Euclidian geometry and the accepted conceptions of space have got to be modified if the general theory of relativity is true. The same thing holds for the conception of time, for a clock in the periphery of the disk would go slower referred to a system than a clock at the center of the disk, as was also shown in the first lecture.

The Daily Princetonian, May 11, pp. 1–2.

Einstein Predicts Shift in Sun Lines

Change Toward Red End of Spectrum Would Prove Atomic Vibrations Vary— Relativity Rests on Tests—Similar Clocks on Sun and Earth Would Keep Different Time, He Is Confident

Special to *The New York Times.*

PRINCETON, N.J., May 10.—The interesting statement was made by Professor Albert Einstein in discussing his theory of general relativity here today that experiments are now being made to determine whether or not there is a shift in spectral lines coming from elements in the sun, which would show that there is a difference in the rate of vibration of the chemical atoms in the sun and the earth, or, in other words, that a clock on the sun would move at a different rate from a similar clock on the earth.

This change of time is a necessary conditional result of the laws of relativity, and he is so confident that experiments will prove his point that he bases the truth of his whole theory on this difference in time. If this condition does not exist he will be forced to admit that his whole theory is wrong, but as it has been verified in two other instances, one of which he predicted before the condition was actually observed, he is not worrying much about the outcome of the present inquiry.

Experiments to show whether or not there are such changes in the spectral lines are being made by Professor Charles Edward St. John of the Mt. Wilson Observatory. Some of his work seems to indicate that Professor Einstein's conclusions are not justified by observation, but his work has not yet reached a point where he is willing to announce a definite result.

Must Change Our Ideas of Space

Professor Einstein allowed yesterday that the special theory of relativity is inconsistent with Euclidean geometry by his favorite illustration of the measuring sticks, this time used at the circumference of a rotating circular disk, and also showed that conceptions of space have got to be modified if the general theory of relativity is true. A résumé of his talk was given in English by Professor E. P. Adams. . . .

[Here follows a part of the résumé that appears in the next article.]

The New York Times, May 11, p. 17.

Einstein Upsets Euclidian Laws

Theory Clears Up Principle of Inertia—Professor, in Second Princeton Lecture, Seeks Proofs in Sun and Mercury.

(Special Dispatch to the *Evening Post*)

PRINCETON, May 11.—Having in his first lecture discussed the special theory of relativity and having postulated that all inertial systems—that is, those which move relatively to each other with uniform rectilinear motion—are equivalent for describing physical phenomena, Dr. Einstein, in his series of five talks at Princeton, discussed yesterday in the second discourse particular problems involved and the important consequences resulting from acceptance of this doctrine.

After Dr. Einstein had delivered his address in German upon "Generalities on the Theory of Relativity," a continuation of Monday's topic, Prof. Edwin Plimpton Adams of the department of physics at Princeton again made a résumé in English, giving the substance of Dr. Einstein's discourse as follows:

"The lecture yesterday had to do with the special theory of relativity, and the conclusion which was arrived at at the end was that all inertial systems, that is all systems which move relatively to each other with uniform rectilinear motion are equivalent for describing physical phenomena. The special theory of relativity depends upon one hypothesis, that the velocity with which light waves travel is the same in every case when referred to an inertial system, that is, systems which move relatively to each other with velocities which are uniform and rectilinear.

"Now, the striking thing about this is that it appears to hold only for a particular type of systems; that is, systems which move with a particular type of motion relatively to each other. The question that we have to ask ourselves is whether the natural laws are such that there should be such type of motion, whether it should not be so with all kinds of motion—systems moving with perfectly arbitrary velocities with respect to each other should not be wholly equivalent as far as describing physical phenomena.

"Now, there is one thing which has been known for a long time which helped a great deal in getting at what we call the general principle of relativity. This is the principle of inertia. We know that in respect to any one of our inertial systems a single mass point, if it is far enough away from another body, moves with uniform velocity in a straight line. But this principle of inertia certainly does not hold when we refer the motion of a body to a system moving in arbitrary motion, say, in particular, a system which is moving with a motion of rotation. In that case, instead of the motion of the particle that we considered first being in a straight line it may in certain cases be in a very complicated curve.

"Now, if all systems, no matter what their type of motion, are to be equivalent for describing physical phenomena, then we have got to change the whole foundation of our mechanics. The principal thing which helps in this change is the known law that the inertia of a body is proportional to the gravitation of that body, or, in other words, to its weight. We measure the inertia of a body by finding out how much acceleration a definite force gives it. We measure the weight of a body, or the gravitational mass of a body, by finding out how much acceleration the gravitational field of the earth gives it.

Clears Up Laws of Inertia

"It is known that these two masses, the inert mass and the gravitational mass, are strictly proportional to each other, and that they do not change for different bodies. In other words, it makes no difference at all whether we speak of the inert mass of a body or whether we speak of the gravitational mass of the body. Just exactly what the significance of this fact is has been a puzzle for a long time, which, however, the general theory of relativity quickly clears up.

"Suppose we consider a system of coördinates which we will suppose to be absolutely at rest. And let us assume that there is no gravitational field present in this system. Consider another system which we will say is moving upwards with accelerated motion. Now, if we have a mass particle in our first system which was at rest, then, in the simplest case, we would suppose that our mass particle was at rest at one instant and therefore, according to our assumption, remains at rest forever. That same particle, however, referred to our second system will be found to have a uniform acceleration downwards.

"The principle of equivalence states this, that it makes absolutely no difference whether we say that our first system is at rest and the second system is uniformly accelerated, or whether we say that we have in place of a uniformly accelerated system a gravitational field present. This principle of equivalence is the foundation of the general principle of relativity.

"Now, the problem that has got to be solved is this: it must be possible if this principle is to hold with any generality to be able to describe any kind of a gravitational field in terms of the motion of a particular system of coördinates. That, however, is not the way in which the general principle has been developed. There is this difficulty which we meet with at once. It can be shown easily that if this general principle of relativity is to hold, then our ordinary concepts of space and time have got to be profoundly modified. You can put it in another way. We can show that the Euclidian geometry will not hold if the general principle of relativity is to hold.

"So far we have spoken only of the fact that Euclidian geometry and our concepts of space have got to be modified if the general theory of relativity is true.

"The same thing holds for our conception of time, for we also saw yesterday that a clock on the periphery of our disk would go slower referred to a system at rest than a clock at the centre of the disk. This is due to the fact that the periphery is in motion with reference to the system of reference and the centre is at rest, so that also our conception of time has got to be modified if the general theory of relativity is true.

"The problem in the general theory of relativity is to find an expression for the laws of natural phenomena so that the principle of equivalence, for one thing, will hold. That, however, we have just seen cannot be done if we are going to hold to Euclidian geometry and to our ordinary concept of time. An analogy here, however, helps a great deal in understanding exactly the mathematical method which can be employed to find out the form of the laws of nature in the general theory of relativity.

"If we consider a surface—of course a surface may be a plane, but take any kind of a surface, a curved surface—if we want to study the geometry of a curved surface we have got to have different methods from those which are used in studying the geometry of a plane surface. In particular one method of describing the position of a point on a plane surface would be exactly the method used in the illustration of the disk. We could lay out a certain number of small measuring rods in two directions at right angles to each other, and then any point in a plane would be given by the number of measuring rods going in one direction and the number of rods that we have to go in another direction. This method, however, is not directly applicable in case we are using curved surfaces. In other words, the geometry of a curved surface is wholly different from the geometry of a plane surface. All we can say is that the geometry of a curved surface is non-Euclidian geometry, while the geometry of a plane surface would be Euclidian geometry.

Laws Must Permit Transfer

"We can, however—and this is the method introduced by Gauss to describe the position of a point in a curved surface—suppose that we draw on this surface two systems of curves in any way whatever—two wholly distinct systems of curves—and number these two systems of curves. Then any point on a curved surface would be located by the two numbers which express the two curves of our systems which intersect at that point. And that is the method that has to be used in describing positions of points on curved surfaces.

"This is merely an analogy, of course, to what is necessary in the general theory of relativity. We have got to be able to express all physical phenomena in such a way that the laws of the physical phenomena shall be unchanged when we transform from one system of these Gaussian coördinates in the four dimensional space involving the ordinary space coördinates to any other Gaussian system of coördinates. And the exact methods in which that can be done Prof. Einstein is going to take up in the succeeding lectures.

"Just as a summary of what will come out from that, we can say merely this: When we want to express the length of a small element on a curved surface that is done by means of certain quantities describing these particular systems of curves that we have taken in the curved surfaces, and there are certain coefficients which are involved in the expression for a line element on our generalized four-dimensional surface as quantities which of course vary from point to point, but they are also important in describing or in characterizing the gravitational field which is present. So that what needs to be done is to express physical laws in the particular form of these Gaussian coördinates. So far it has been done principally for the laws of gravitation, and that is one particular additional hypothesis which has to be introduced to derive those laws, the hypothesis that only certain differential coefficients can come into the expressions.

"Now, the consequences of the general theory of relativity. First of all one thing is necessary. If the general theory is to be true it must result that the laws of gravitation deduced from the general theory of relativity are at least as accurate when tested astronomically, as the laws derived from Newton's laws of motion and Newton's laws of gravity. And this conclusion has been completely verified, that the difference between the numerical results of the general theory of relativity are remarkably nearly the same as the numerical agreement in the case of Newton's laws. There is one difficulty, however, which the general theory of relativity removes which has been a difficulty for over a hundred years in the Newtonian theory of gravitation. That is in respect to the motion of the planet Mercury.

Shows Motion of Mercury

"Instead of the planet Mercury describing an ellipse which remains a fixed ellipse with respect to the fixed stars, the axis of this ellipse continually rotates in the direction of the rotation of the planet about the sun. This rotation amounts to about forty-two seconds of an arc in one hundred years, and has been for a long time a very puzzling difficulty in the ordinary dynamic theory. The general theory of relativity predicts in such a case a revolution of almost exactly this same amount, forty-two seconds in one hundred years, the same amount that has been observed, and that, of course, is a remarkable result of the general theory.

"There is one other result, however, of the general theory which is even more remarkable for the reason that the result was predicted before the experimental fact was determined, and that has to do with the bending of light when it travels near a heavy body, as in particular the light from a star which can be observed at the time of a total eclipse of the sun. When this light passes near to the sun there is an extremely slight bending of the light, but one which was found by the two English expeditions sent out to agree within about 10 per cent of the amount predicted by the theory. And that, of course, is really the more remarkable result of the general theory of relativity, in that it has predicted a result which was not previously known.

There is a third consequence of considerable importance, because it has not yet been verified. That has to do with the change of time. If we had a clock in the sun, being in a different gravitational field from a clock on the earth, we would expect from the general theory of relativity that the two clocks, the one on the sun and the one on the earth, even if they were exactly alike, would go at different rates. We have such clocks in the vibrations, whatever they are, which take place in the chemical atoms, and these clocks give us these indications by the spectral lines emitted from them.

"The consequence predicted by the general theory of relativity is that all spectral lines coming from elements in the sun should be shifted extremely slightly towards the red end of the spectrum. Whether or not this effect exists is still a matter of controversy. And the extremely interesting remark that Prof. Einstein made near the end was that if it is definitely proved that this effect does not exist, then his whole general theory of relativity falls down. And his confidence in the theory is apparently such as to lead him to believe that such an effect really will be found to exist.

"That, then, is very briefly the content of the general theory of relativity, and perhaps the most important consequence of it is the fact that our whole concept of space and time has got to be altered. We can no longer think of space, time, and matter as independent concepts, but they are interwoven upon each other."

The New York Evening Post, May 11, p. 7.

Wednesday, May 11

Einstein Praises American Colleges

Thinks Research Work Greater Than Abroad—Scientist Outlines Special Theory of Relativity and Answers Critics

(Special Dispatch to the *Evening Post*)

PRINCETON, May 12.—Before a small group of mathematicians, physicists, astronomers, and graduate scholars in sciences, Dr. Albert Einstein, in the third lecture of his series delivered at Princeton yesterday, delved into the intricacies and technicalities of his subject. The discourse was devoted to the mathematical formulation of the special theory of relativity in the form given it by Minkowski.

After the colloquium which followed the lecture, Dr. Einstein, as he walked along the street with a band of faithful satellites and admirers at his heels, made some general observations upon scientific research in this country and in Europe. He said in substance, as translated from the German:

"I am particularly pleased with the amount of study being done in America in connection with this theory and with other scientific matters. Because of the unfortunate conditions in Europe at present, the chances for arriving at important discoveries in this country are now far greater than in the intellectual centres of the Old World.

"Due to the economic conditions in Europe, scientific research at the universities is at a standstill. All the educators and scientists there are using their heart, head, and hand for whatever purpose can gain for them the means of subsistence.

Praises American Colleges

"The result is that scholarly investigation has had to be abandoned, and it will be a good many years before the nations will recover from the present economic situation so that the university men can resume their researches without worrying over the problem of making both ends meet, financially."

Dr. Einstein was especially impressed by the way the students on the Princeton campus live under the dormitory system, thus making, as he put it, for their "social well being." The average man, Dr. Einstein thinks, is probably prevented from doing his best work possible in studies by the attractive environment and the opportunities for enjoyment afforded in most American colleges he has seen. On the other hand, the visiting scientist declared that the ones really eager in the pursuit of wisdom, could do better work on the college campuses of this country than those among European surroundings.

The fourth lecture of the series will deal with "The General Theory of Relativity and the Gravitational Field," and in the concluding talk Dr. Einstein will engage in "Cosmological Speculations." The ground covered by the visiting physicist in his discourse yesterday upon "The Special Theory of Relativity," was thus summarized by Prof. Edwin Plimpton Adams of the Princeton faculty.

The third lecture was devoted to the mathematical formulation of the special theory of relativity in the form given it by Minkowski. General theorems regarding the properties of four-dimensional vectors and tensors were stated and the Maxwell-Lorentz equations of the electro-magnetic field were written in terms of these quantities.

"It was shown how the ordinary electro-magnetic forces due to motions in the electro-magnetic field followed as an immediate consequence of the special theory of relativity. Finally the equivalence of mass and energy resulting from this theory was brought out, as a consequence of which the two laws of conservation of mass and energy become welded into a single law.

"Following the lecture a colloquium was held for the discussion of difficulties. Dr. Silberstein, scientific adviser to the Eastman Kodak Company, acted as interpreter. The principal topic of discussion related to Prof. Einstein's illustration of a rotating disk. Prof. Einstein removed the difficulties which had been felt by remarking that the objective existence of the rotating disk was not at all necessary for the purpose of the illustration."

While the highly trained mathematicians are thus wrestling with the complicated formulae and conceptions involved in the advanced aspects of the theory, Prof. Luther P. Eisenhart of the department of mathematics at

Princeton has given an answer to the query whether such speculations are "much ado about nothin."

"The question has been raised whether the Einstein theory is any good after all," said Prof. Eisenhart. "The purpose of any physical theory is to give a mathematical formulation which agrees with experiment and from which predictions can be made concerning phenomena which have not been tested by experiment.

"As further experimental results are obtained it may be necessary to change the physical interpretation which the genius of Einstein gave to a mathematical structure developed by a group of brilliant geometers from the time of Riemann. But Einstein's conception of the relation between geometry and physics is likely to exert a lasting effect upon theories concerning the physical world.

Dr. Brush's Contradictions

"Recently Dr. Brush announced that he had shown by experiment that gravitation acts differently upon subjects of different physical nature. Both the Newtonian and Einstein theories of gravitation are based upon the conception that this is not the case. Dr. Brush made no reference to the experiments of Baron Eötvös of Hungary in 1890, by which he showed that the action of gravitation is independent of the physical nature of a body.

"These experiments purport to be much more accurate that those of Dr. Brush. They have been accepted as conclusive by physicists, and certainly must be shown to be incorrect before Dr. Brush's results can be accepted."

One fundamental difference between the Newtonian idea of gravitation and that sponsored by Dr. Einstein is pointed out as follows in non-technical language by Prof. Eisenhart:

"According to Newton's laws of gravitation, two bodies attract one another with a force proportional to the product of their masses and inversely as the square of the distance between the bodies; it is based upon the principles of absolute space and time, and deals with action at a distance.

"The Einstein law is based upon the principle that the character of physical space is determined by the presence and distribution of matter, that it is not Euclidian, and that action takes place in the neighborhood of a body, just as physicists from the time of Faraday have believed that electrical action occurs.

"By the consideration of certain mechanical problems, Einstein came to the conclusion that in general there is no particular system of time and space coördinates which are fundamental. Consequently he has formulated his theory in times [terms] of any system whatever. This necessarily makes the mathematical formulation quite involved.

"However, in the consideration of special problems, such as planetary motion, the equations are of the same order of difficulty as in the Newtonian mechanics. In fact, the equations of planetary motion in the two theories differ so little that only in the case of the motion of Mercury is the difference detectable by experimental methods now available. Likewise in the treatment of the usual mechanical problems met with on the earth the Newtonian mechanics can be employed. For some time it has been recognized that this is not the case with electro-dynamics as formulated by Maxwell. However, this theory has been coördinated with the general theory of relativity."

Discussing the vagaries of the planet Mercury and their importance in the doctrines as developed by Einstein, Prof. Eisenhart said:

"His prediction that a ray of light passing near the sun would be bent was a by-product of the theory of gravitation which he had been developing. He had shown that this theory accounted for the discrepancy in the motion of the planet Mercury which had been known for 200 years.

"This was an interesting fact, but there had been other explanations of this phenomenon. Recently we have been told that it would be accounted for by a certain distribution of finely divided matter in space, but this requires experimental verification. Also it is possible to make a change in Newton's laws which will account for the discrepancy. But a verified prediction gives to a theory a claim to be reckoned with."

The essence of the findings by Einstein is, as Prof. Eisenhart points out, that "the magnitudes of observed space intervals and time intervals are relative to the observer. These conclusions are based upon the hypothesis that the velocity of light is independent of the velocity of the source of light; this assumption seems to agree with experiment. These ideas form the basis of the so-called restricted principle of relativity. The general relativity of Einstein represents an extension of these ideas so as to include curvilinear motion and gravitation."

The New York Evening Post, May 12, p. 7.

Scientist Discusses College Work Here

Einstein Comments on Social Aspects of Life—Science Hampered in Europe—Technicalities Taken Up in Yesterday's Lecture—Scientist Reveals Mathematical Formation of Theory of Relativity in Detail

In a statement to a representative of the *Princetonian* after the lecture yesterday, Dr. Einstein expressed his opinion in regard to scientific progress and the betterment of work in college. Due to the present economic conditions in Europe, scientific research is more or less at a standstill. All educators are in straitened circumstances, using hands and head in an effort to maintain a bare existence. For this reason, it will be a good many years before normal conditions will again prevail and they will be assured of regular income. Therefore, since the well-being of the teaching profession here is on a more certain basis and better chances in America have made investigation possible, more progress has been made in this country.

Speaks of Dormitory Life

Professor Einstein was struck with the way in which the students live together in the dormitories at Princeton, as well as the social aspects and advantages of their life as a whole. While he considered that the numerous social activities and demands of friendship would prevent the average man from studying and offer a difficult distraction, he believed that anyone really interested in study could do better in the dormitory environment than if the students all lived at home, as in many of the German universities.

In his lecture yesterday, Dr. Einstein discussed the "Special Theory of Relativity," in which he took up more of the intricate technicalities of his subject. Today at 4:15 in McCosh 50, he will discuss the "General Theory of Relativity and Gravitation."

A Consequence of Theory

In his second lecture, Dr. Einstein showed some of the important consequences of his theory. If his general theory is true, then it must follow that the laws of gravitation deduced from that theory are as accurate, when tested astronomically, as Newton's laws.

A remarkable consequence of the theory was the prediction in regard to the bending of a ray of light from a star passing near the sun. The amount of bending predicted agreed within ten per cent to the amount ascertained by two English expeditions.

Yesterday he turned from these broader aspects of the relativity principle to discuss its mathematical formation. It was shown how the ordinary electro-magnetic forces due to motions in the electro-magnetic field followed as an immediate consequence of the special theory of relativity. Following the lecture a colloquium was held for the discussion of difficulties. Dr. Silberstein, scientific adviser to the Eastman Kodak Company, acted as interpreter.

The Daily Princetonian, May 12, pp. 1–2.

Boy of Sixteen Understands Einstein

Jewish Lad of Brooklyn Amazes His Teachers

Only twelve men in the world, according to experts, understand the Einstein theory of relativity, but Brooklyn teachers are confident the borough has a thirteenth in Theodore Herberg of 703 Georgia Avenue, a boy of sixteen, who is a pupil at Boys High School.

Young Herberg, who reads books on physics and mathematics as other boys read "Treasure Island," gave a practical demonstration of his knowledge before the faculty members which left the teachers gasping.

First young Herberg showed certain calculations as revealed by the Michaelson Morley experiments with relation to speed. Before the teachers caught their breath, he was explaining the Lorenz Fitzgerald contraction formulas for bodies moving through space at a great velocity, demonstrating in accordance that a physical body of a known length would contract to a point if the body were to be moved with the velocity of light.

In a playful manner, he set out to prove some of the consequences of relativity, one of the demonstrations going something like this:

"A ship is sailing from North to South, the wind is blowing from the South, the earth is rolling round on its axis from West to East, the captain is walking across the deck of the ship from East to West, while a monkey runs up to his shoulder" . . .

At this point the reporter pressed a shaking hand to a fevered brow, and the rest was lost.

As far as could be gathered, however, the question decided was the direction in which the monkey was proceeding in relation to all the other factors.

Herberg explained.

Shaken and bewildered, the teachers clung desperately to their chairs, while the young scientist adjusted his glasses, and smilingly asked for questions.

Relativity is only one of the boy's hobbies. He is also editor of a Latin magazine published at the school. He is a sketch artist. He leads the school in physics, mathematics and Latin, and though he could have graduated a year ago, has chosen to remain at school to study "Greek and a few other subjects." Otherwise, he is quite human, for when a reporter went to see him at the Boys High School, it was in the gymnasium that he was finally run to earth.

Yidishes Togeblat—The Jewish Daily News, May 12.

> Young Herberg did not outgrow his interest in mathematics: he wrote at least four textbooks on math and geometry in the 1930s. He worked in Pittsfield, Massachusetts, first as a teacher of math at the high school, then as director of curriculum and testing, staying at the school for more than 70 years in all. A middle school and a library preserve his name there.

Future of Science Here, Says Einstein

America Must Keep Up Research While Europe Wallows in War's Aftermath—Praises U.S. Colleges

PRINCETON, May 12. (Special).—The future of science rests for the present with the United States, according to Dr. Albert Einstein, the "apostle of relativity."

Because of economic exhaustion as a result of the war, all scientific research in Europe is at a standstill, he said.

He likes the American college system, the visitor said.

"I am particularly struck," he said, "by the way the students live at Princeton; the social aspect as well as the social well-being. While the life in dormitories and other social aspects of American college life would prevent the average man from studying, any man who wanted to study could probably do better in the American type of college than in Europe, where the students live in the cities."

Princeton, which has been lauded by Dr. Einstein as one of the first colleges to appreciate and study his theory, welcomed the scientist Monday in McCosh Hall with a crowd that filled the hall to the doors. On the second day the auditorium was half-filled. On the third day the discussion was moved to a small class room. And as the explanation progressed through its one hour and a half length, in groups of two and three the scientists dropped out of the discussion and out of the room.

Princeton savants in deep scientific communication with Dr. Einstein, are discussing this week exactly the same things that Zeno and Pythagoras discussed 2,500 years ago.

Is there any difference between rest and motion? Is length absolute? Do time and space have real existence?

Those questions were discussed by the Greek philosophers in their temples and in the cities of Asia Minor in the days before Christianity.

So, point out certain scientists here, the Einstein theory has revived a question 2,500 years old. Zeno argued against the reality of space. Pythagoras, in his philosophy of the limited, or definite, forms, and the unlimited, or space, attempted to bring the two together in harmony. Such are the efforts of Einstein.

To understand the new theory of gravitation by Einstein, it is said one must have a knowledge of the theory of variants and of the calculus of variations.

"Does the theory include a sun as the centre of a universe of worlds, each having suns similar to ours, and all controlled by one central, stationary sun?"

The question was read to Dr. Einstein several times. He laughed.

"No," was the answer, but he hastened to explain further that it would take him all night and probably part of the next day to explain his answer with its qualifications.

The Evening Bulletin (Philadelphia), May 12, p. 20.

Thursday, May 12

Declares Einstein Is Inconsistent

Princeton Professor Discusses Philosophical Theory—Fourth of Series of Lectures on Relativity Delivered Before Scientists

(Special Dispatch to the *Evening Post*)

PRINCETON, May 13.—Dr. Albert Einstein delivered the fourth lecture of his series of five yesterday, devoting his attentions to the mathematical development of the general theory of relativity, with special reference to the gravitational field. Attendance was confined to mathematicians, physicists, and graduate scholars.

The fifth and last of the series will be upon "Cosmological Speculations." As this topic is more general than the two preceding ones, and more nearly similar to the subject-matter of the opening lectures, it is expected that not only the specialists but also many laymen interested in the subject will attend.

The following short résumé of yesterday's discourses was made by Prof. E. P. Adams of the Princeton faculty, who has from day to day given in English the substance of Dr. Einstein's lectures.

"The fourth lecture was devoted to the mathematical development of the general theory of relativity. Beginning with the analogy of Gauss's theory of curved surfaces, Prof. Einstein showed how the methods of the absolute differential calculus developed by Riemann and others led to a formulation of the principle of equivalence, by means of which the gravitational field could be interpreted in terms of the invariant properties of four dimensional space.

"He finally showed the form that the equations of motion of a material point take in a gravitational field, and also how the path of light rays in a gravitational field can be determined.

"He announced that the final lecture of the series would be devoted to conclusions with respect to the universe as a whole which may be drawn from the general theory of relativity."

Discuss Philosophical Aspects

To those who have wondered whether the acceptance of the Einstein principles would make it necessary to alter all the ordinary beliefs concerning the universe and human life, this statement of Prof. E. G. Spaulding of the department of philosophy at Princeton should be reassuring. Prof. Spaulding, who has made a special study of the new theory in its relation to his field of research, said yesterday:

"The Einstein theories have no bearing at all upon problems of ethics, problems of psychology, problems of theology, or, in general, upon the most important ones of philosophy." This new theory therefore does not upset the laws of morality, and the principle of relativity, as Prof. Spaulding points out, does not apply to evaluations of goodness.

But it would be a mistake, Prof. Spaulding thinks, to assume that the researches of the Swiss physicist have no bearing at all upon philosophical questions. For instance, if the Einstein conception of a finite physical universe be accepted, one of the problems which has engaged philosophers for centuries is automatically solved.

"Furthermore," said Prof. Spaulding, "this view that space, time, matter, and energy are interwoven and interdependent would modify certain philosophical positions which regard time and space as independent and indeed 'a priori' categories, as, for instance, in the Kantian system.

"But there is nothing in the Einstein theories to settle the age-long issue between two such philosophies as idealism and realism. Idealism has always been able to take up into itself all the detailed scientific data and hypotheses concerning nature.

"Throughout his philosophy Einstein is very much of a pragmatist, that is, for him the test of truth is the practical consequences; and the possession of truth is not an end in itself but only a means to a larger practical purpose. This pragmatic attitude is exemplified when he shows that there is no such thing as the absolute truth with respect to geometrical propositions, but that what geometry one uses depends upon one's purpose."

Prof. Spaulding regards the Einstein principles as significant, not only because of their importance in the spheres of physics and mathematics, but because they demonstrate that science as a whole, which has prided itself upon dealing with facts in an unbiassed manner, unprejudiced by the speculations rife in other branches of knowledge, is not free from the fault of proceeding upon preconceived notions without sufficient experimental verification.

"It has been thought that the Newtonian mechanics were a criterion of absolute knowledge, but if the Einstein theories are accepted, the conclusion can be reached that the law of progress of science is one of repeated correction and ever closer approximation, but not one of definite, absolutely true knowledge. These new doctrines make it plain that there has been dogmatism in science as well as in theology. Certain philosophies have arisen in the past as protests against scientific dogmatism.

"The Einstein theories show how crude was the old materialism which conceived of everything as consisting of masses in motion. They demonstrate that the real structure of things may be that which seems inconceivable to the ordinary minds, and even to the usual scientific minds. In general, these new tenets should be regarded as making for a much greater openness of mind and catholicity of view on the part of scientists."

Points Out Inconsistencies

Prof. Spaulding then said that, for himself, he was willing to accept, with reservations, the propositions laid down by Dr. Einstein, and to work out some philosophical problems on the basis of new conceptions brought forward by the principle of relativity. But he pointed out certain parts of the theory which seem to contradict the general tenet.

"Thus, up to a certain point the Einstein theories seem to be consistent and convincing, but beyond that point some contradictions appear to present themselves. For example, the relativity which runs through his theory seems to be limited at a certain point by something absolute." (He here referred to the problem of the rotating discs, a phase of the theory too involved for ordinary terminology and conceptions.)

Another apparent contradiction, or at least inconsistency, is thus pointed out by Dr. Spaulding:

"The Einstein principle of the change of length of a measuring rod with the motion of the rod parallel to its length is not a change if the length is determined by another smaller rod carried with the first. It is a change only in relation to something which seemingly would be an absolute character of space."

In conclusion Prof. Spaulding said:

"Einstein, despite the shifting coordinates and transformation of the direction and velocity of motion, and despite the shifting size and time involved in his theory, nevertheless insists that if there are laws of nature they must be invariant and universal."

The New York Evening Post, May 13, p. 7.

The three paragraphs of Adams's account were also published in the (Philadelphia) *Public Ledger,* May 13, p. 1, and in the *Boston Globe,* May 13, p. 12.

Spaulding's view is also summarized by the (Philadelphia) *Evening Bulletin,* May 13, p. 15, with the following paragraphs added:

Professor J. Duncan Spaeth, of the English Department and coach of the Princeton crew, believes that the theory offers possibilities for application to every day life.

"For instance," said Professor Spaeth, "here I am considered a modernist, and when I am on the Pacific coast I am considered a classicist. It all depends upon the angle from which you look at a man."

On politics and human relations, he believes, the theory of relativity could be applied in our appraisement of our fellow-citizens.

Last Einstein Lecture Given This Afternoon

Famous Speaker to Talk on "Cosmological Speculations" in McCosh 50 at 4:15

In his lecture at 4:15 yesterday, Dr. Einstein took up the technical side of his general theory of relativity, comparing it with the technical side of his *special* theory of relativity, discussed the day before. The speaker used extremely complicated formulae in his attempt to explain the mathematical formation in a specific way. The blackboard was covered with abstruse diagrams which only a trained mathematician could follow.

Terminates Visit

Today at 4:15, Dr. Einstein will end his series of talks at Princeton with the most famous of all his lectures, on "Cosmological Speculations." This will deal with the subject in its broader aspects and should be of great interest to the average student. Dr. Einstein will advance some startling conceptions as to the results of his theory. This will terminate the scientist's visit at Princeton.

The talk yesterday again took up the fact that the predictions of the theory of relativity were incompatible with Euclidean geometry, and it was shown that the latter geometry was inadequate to explain the laws of many natural phenomena. Fundamental changes in the accepted conceptions of time and space are necessary; these changes are explained by Dr. Einstein's theory which has opened new fields of truth and progress.

The Daily Princetonian, May 13, p. 1.

Hope Soon to Prove If Einstein Is Right

Terrestrial Optical Experiment to Test Pull of Rotating Earth Upon the Ether— Michelson Ready to Try It—New Theory Evolved by Dr. Ludwik Silberstein Holds Interest of Savants

Special to *The New York Times.*

PRINCETON, N.J., May 12.—A proposal for an experiment which may prove Einstein's theory of relativity to be all wrong has been placed before scientific men here to attend Professor Einstein's lectures, and it has aroused the greatest interest. This is to test the pull of the rotating earth upon the ether to learn whether there is a drag, whole or partial, and it has several possible results, the most important of which is its effect on the theory of relativity.

So important is the experiment judged to be by those who have learned of it that Professor Albert A. Michelson, the man who made experiments to determine the passage of the earth through the ether—the failure of which to give a positive result led to the discovery by Einstein of his laws of relativity—and who recently discovered a way to measure the diameter of fixed stars, has offered to perform the experiment itself. Professor Einstein was informed of it three days ago, and at first was inclined to doubt that it would have any bearing on his theory, but, after thinking it over, has decided that it is a new and practical way of testing his theory, and has described it as "wonderful."

The experiment has been evolved after months of calculation by Dr. Ludwik Silberstein, of the research laboratories of the Eastman Kodak Co, in Rochester. He came here to talk over his theory with Professor Einstein and has had several conversations with him which have aroused intense interest among other visiting scientists. As outlined by Dr. Silberstein his theory is briefly this:

What New Theory Means

If a beam of light is sent around a closed circuit in one direction and at the same time in the other direction, the time taken by the light in both cases is equal, provided that the experiment is performed by means of an apparatus fixed in the luminiferous medium, or ether. If, however, the apparatus is rotating with the earth with respect to the ether, then the times taken by the light to cover the circuit in one and the opposite directions will differ.

The lagging behind of one relative to the other will be greater the greater the velocity and the greater the area enclosed within the circuit. To give a numerical example: If the enclosed area of a horizontal circuit is one square kilometer and if green light be used, and the experiment is performed at the latitude of Chicago, and if the angular velocity relative to the ether is 360 degrees each twenty-four hours, then the retardation will be as great as 1.4 periods of oscillation of the light used.

By modern interferometric methods, the method used by Professor Michaelson in determining the size of the star Betelgeuse, in the hands of skilled investigators, .001 of a period can be measured. The retardation of a period means visually the shift of certain interference patterns by the whole fringe widths, and it is possible to measure exactly .001 of a shift.

The shift is ascertained by sending the split light rays around the circuit of an equilateral triangle. At one corner of the triangle is placed a semi-silvered plate of glass, whose function it is to split the light beam into two partial beams. The other two corners are occupied by mirrors. This triangle may enclose an area one square kilometer in extent although Dr. Silberstein hopes it will be possible to perform the experiment with a triangle of only .1 of a square kilometer, or even less.

Light Rays to Be Used

The light beam reflected from the silvered portion of the plate of glass moves in a curve resembling a parabola from point to point on the inside of the triangle, and against the rotation of the earth. This is the ray that is first sent by reflection on shutting off the transparent portion of the glass. The second ray goes in the opposite direction through the unsilvered portion of the glass, and also moves in a curve almost parabola from point to point on the outside of the triangle. The difficulty arising from the impossibility of inverting the direction of the earth's rotation will be met by an ingenious technical device invented by Prof. Michelson.

The point of interference of these two light rays is where the delicate measurements which will detect the value of the experiment will be made. The ray which travels against the earth's motion should go around fastest, Dr. Silberstein explained, and the second ray which travels on the outside of the triangle in the direction of the earth's motion should go slower. By first covering the transparent portion and then uncovering it so as to let the second ray travel around the triangle a shift of the interference patterns at the point of intersection should be apparent.

This consequence of the other theory is obvious and already known, Dr. Silberstein said, and he had drawn attention to it in a paper recently read in Chicago. He worked out the shape of the rays, which are more rigorously spiral shaped than the usual parabola. Moreover he has worked out the theory of this terrestrial experiment on Einstein's theory of general relativity, and found that in this latter theory the shift effect should have necessarily the full value of 1.4 fringe widths in an experiment performed with green light at the latitude of Chicago.

What It Will Prove

Based on the ether theory the effect should be either equal to this full value, if there is no dragging of the ether by the spinning earth, and no effect at all if there is a full drag. Finally there would be only a fraction of the full effect if there is a partial dragging of the ether by the spinning earth.

If, therefore, the experiment which Professor Michelson will perform gives a full value of the shift, this will harmonize with the general relativity theory as well as with the other theory, but if the effect is nil, or only a fraction of the full shift of 1.4 per square kilometer, it will be "a death blow to the relativity theory," although compatible with the ether theory, testifying simply to a partial drag.

When Dr. Silberstein first outlined his idea to Professor Einstein the latter was doubtful of its application, but after thinking it over a day or two, told Dr. Silberstein that he appreciated the soundness of the reasoning on which it was based, and said it was of the highest interest. He showed the greatest impartiality, and said he would gladly recognize a fractional shift as a blow to his theory, and at the same time enjoy the demonstration of the novel phenomena. However, both Dr. Silberstein and Professor Einstein believe that the full shift effect will be shown.

Dr. Silberstein warned against any confusion of this experiment with the Michelson-Morley experiment which led to the theory of relativity, but there is this similarity in that they both used light rays for which the ether is the medium to detect motion of the earth in the ether. In the Michelson-Morley experiment the attempt was made to learn the translational motion of the earth, whereas in this experiment the attempt will be made to determine what effect the spinning motion of the earth has upon the ether.

In the Michelson-Morley experiment the light ray was split and one part sent forward in the direction of the earth's motion, and the other sent to a mirror on the side. It was argued that by sending light signals through equal distances in different directions the light to the side should return before the light sent on a chase after the mirror which is moving ahead with the speed of the earth.

"If one had gained on the other by even a fraction of the time of vibration of a single light wave the fact could be detected," wrote Professor Henry Morris Russell, the Princeton astronomer, in speaking of this experiment, "and the waves which we ordinarily call light vibrate at the rate of about six hundred thousand billion per second.

"Michelson and Morley tried their experiment, and in place of the easily measurable results which they anticipated, they got nothing. The light waves came back over the two paths in exactly the same interval of time."

When this came to Einstein's attention and he studied it for a time he came to the conclusion that it was not possible by physical experiment to detect the existence of absolute straightahead motion, and laid down the principle that only the relative motions of bodies in the universe can be studied.

And as the original Michelson-Morley experiment gave rise to the theory of relativity, Dr. Silberstein explained, so the experiment to determine the pull on the ether by the spinning earth may either corroborate the Einstein theory or destroy it.

The New York Times, May 13, p. 7.

> There is no unambiguous historical evidence that the Michelson-Morley experiment inspired Einstein to develop special relativity. Here you see an example of how a logical connection is considered a historical one—a mistake often seen with scientists.

Dr. Einstein was a special guest

at the Nassau Club Luncheon on May 11th, and was also the dinner guest of the Departments of Physics and Mathematics at the Nassau Club on the 12th. Many of the visiting scientists were also guests of the club during their visit. At the Club Luncheon Dr. Einstein was presented to the assembled members and guests in German by Dr. Charles Browne '96, who presides at these interesting weekly meetings. On this occasion the Club also had the pleasure of hearing from Dr. M. W. Jacobus '77 of the Board of Trustees, who spoke on the curriculum, and from Mayor Moore of Philadelphia, who told something of the efforts of his administration to remove from his city the stigma of being "corrupt and contented."

The Princeton Alumni Weekly, May 18, pp. 734–37.

Friday, May 13

Einstein Cannot Measure Universe

With Mean Density of Matter Unknown the Problem Is Impossible—Final Princeton Lecture—Universe Called Finite and Yet Infinite Because of Its Curved Nature

Special to *The New York Times.*

PRINCETON, N.J., May 13.—Professor Einstein explained his conception of a finite and yet unlimited universe in the last lecture of a series he has been giving this week at Princeton before scientists who have been working on his theory in this country since it was first announced. Just what the size of the universe is he said could not be determined at present, because it is first necessary to know the mean density of matter in it, and this at present is a quantity of which there is no knowledge.

Professor Einstein's idea of the finite universe is that of a spherical universe of finite extent, but infinite because of its curved nature, as was explained some months ago in *The Times* by Professor L. P. Eisenhart. He conceives the universe as being bent back upon itself much as the mythical snake which swallows its tail, although, of course, there is no way of making a graph of what is a mathematical abstraction. In a summary of Professor Einstein's lecture today Professor Adams of the department of physics said:

"It is a remarkable fact that the general theory of relativity, built up as it is from physical considerations resulting from experiments on the earth, should have anything to say concerning the problem of the universe as a whole. It has generally been thought that the universe is infinite in extent. Telescopes of increasing power have brought more and more distant stars to our vision. If we imagine a sphere of radius very large compared to the mean distance between the stars, our first view is that as we increase the radius of the sphere more and more a definite density of matter in the universe is approached. The astronomer Seeliger first showed that such a view is definitely opposed to the Newtonian law of gravitation, for this view immediately leads to the result that the gravitational field would also increase beyond all limits as we go out toward infinity, and this would mean that the stellar velocities would necessarily increase beyond all limits.

Conflicts with Gravitation

"Thus on the basis of Newton's theory we should have to conclude that the mean density of matter in the universe is zero. This could only be attained by assuming that the universe is an island floating in infinite space free from matter. But this view is wholly unsatisfactory, and Seeliger attempted to reconcile an infinite universe with finite density by assuming that matter of negative density is present in the universe. This assumption involves a departure from Newton's law of gravitation, but no other argument leads to a similar conclusion, and so this is not a satisfactory solution.

"By making slight modifications in his general theory of relativity which do not change any of the other conclusions drawn from it, Professor Einstein shows that a uniform distribution of water [matter] in the universe is possible only in space of constant curvature. In order to be able to form a conception of what is meant by space of constant curvature, we can imagine beings of two dimensions living on a surface so that their space is this surface. If viewed from three dimensions this surface is a plane surface, then the laws of Euclidian geometry would hold. If, however, the surface is a curved surface, for example, the surface of a sphere, and our imaginary beings were unable to go away from the surface, then their space would have a definite curvature at each point and would be non-Euclidian. Such a space for them would be finite, although unbounded. By making measurements on a large enough scale in their space, our imaginary beings would be able to find out whether or not their space was Euclidian.

"And thus in the three dimensional case we can form a conception of how it is that sufficiently large scale measurements will enable us to show that the space we live in has properties different from those of Euclidian space. In order to determine the size of the universe it is necessary to know the mean density of matter in it. But this is a quantity of which we have no knowledge.

The Origin of Matter

"Another remarkable bearing of the general theory of relativity on the physical properties of matter comes from the results of recent researches which indicate that all matter as we know it has an electrical origin. There has, however, always been a very serious difficulty involved in this view. For if we regard a portion of electricity which must be supposed to have a finite although it may be very small volume, the repulsion between elements of that volume requires some unknown force in order that it may hold together. In order to get round this difficulty, Poincaré assumed the existence of a kind of pressure in the universe of sufficient magnitude to balance the electrostatic repulsion between the elements of electricity. Professor Einstein showed that the assumption of pressure throughout the universe is wholly consistent with the general theory of relativity, although it does not follow as a consequence of the theory."

The New York Times, May 14, p. 1.

Dr. Einstein Ends Series of Talks on Relativity

Last Lecture Given in McCosh Yesterday Afternoon—Colloquium Follows Address

In the last lecture of the series, delivered in McCosh 50 yesterday afternoon, Dr. Einstein dealt with the cosmological problem presented by his theory of relativity, and showed that the universe is finite and yet unlimited. Before the lecture, President Hibben spoke briefly of the privilege which Princeton has enjoyed in Professor Einstein's visit. Immediately following the talk, a colloquium was held, at which the distinguished scientist answered questions asked by the mathematicians and physicists who came to Princeton to attend his lectures.

Dean Fine Speaks

In the colloquium some of the details of the theory were discussed. At the end of this, Dean Fine made a short address in which, on behalf of the audience, he thanked Professor Einstein for the great patience and care with which he had explained his theories. Dr. L. A. Bauer, of the Carnegie Institute of Washington then expressed the appreciation of the 283 visiting scientists for the hospitality of Princeton this week. Doctor Bauer was a member of the eclipse expedition to Liberia which obtained the astronomical proof of Einstein's theories. He congratulated Princeton "for having given the time and space for the delivery of these remarkable lectures."

In his lecture, Professor Einstein showed that it was impossible for the material universe to be infinite in extent, for then its gravitational field would be of infinite strength, unless there were some space of negative density. But this is not a satisfactory explanation. A slight modification of the theory of relativity, which does not affect other conclusions from it, shows that a uniform distribution of matter is only possible in space of constant curvature. Such a space is non-Euclidean in its geometry, and lines in it are of infinite length as are the great circles of a sphere. The radius of curvature of such a space could be calculated if the mean density of matter in it were known. But we have no knowledge of this quantity.

The Daily Princetonian, May 14, p. 1.

Apparently the farewell lecture was held again in McCosh 50.

For a full report

of the important series of lectures delivered by Professor Einstein, which were concluded on the 13th, we shall have to await their authorized publication in book form by the Princeton University Press. The daily press reports of the lectures as interpreted by Professor Adams have been unusually good in the difficult circumstances, but such reports, being necessarily brief, cannot be expected to do justice to such an abstruse subject as the theory of relativity, which marks an epoch in man's conception of the universe. The visit of the eminent physicist, the most conspicuous figure in the modern scientific world, has of course been the great intellectual event of the year, or for that matter, of many years. It must have impressed our undergraduates, always busy with their own affairs, to observe the eagerness with which the scientists from other institutions flocked to Princeton to hear Dr. Einstein. It was their only opportunity to hear from the propounder of the new theory while he is in this country, a full account of the results of his many years of research, and during the week there were nearly three hundred visiting scientists in Princeton. They found Professor Einstein extremely modest, open-minded, and ready to discuss with them the new conceptions he has advanced and particularly eager to learn their views and consider their objections to his theory. He gave the impression of the true scholar,—the man eagerly seeking the truth.

The Princeton Alumni Weekly, May 18, pp. 734–37.

> Princeton undergrads were impressed by the eagerness of their professors to learn relativity but the following article more than questions whether the students themselves were eager to follow their professors.

Saturday, May 14

No Student Worry on Einstein Talks

Scientist's Relativity Theory "Shooting Too High for Us," Is Campus Verdict—May Be Popularized

PRINCETON, N.J., May 14 (Special).—

"What's all this Einstein dope about a theory of relativity?"

"Oh" answered the student with the close-fitting skull cap, "it refers to the relationship between space and time and matter."

"Doesn't interest me except when I have to catch a train."

This attitude of the student on the campus of Princeton is likely to be the attitude of the man in the street to what scientists here call a theory that "is the result of the highest achievement of human thought."

Dr. and Mrs. Einstein leave for New York today preparatory to making a trip to Philadelphia early next week, where the scientist will speak for the Zionist cause. They may sail for Europe May 23.

Already Princeton professors have started to polish up "popular lectures" on the subject.

Even members of the Princeton Press Club who won their positions after strong competition confess they are in a "fog."

"What you got on Einstein?" asked the first student of the second.

"I sat in the balcony," is the reply, "but he talked right over my head anyway."

Dr. Einstein in his last lecture in McCosh Hall suggested a means by which the universe might be measured. He said: "In order to determine the size of the universe, it is necessary to determine the mean density of matter in it."

Just as his lay hearers seemed about to hear something they could grasp after a week of deep scientific lectures in German, the scientist added:

"But this is a quality of which we have no knowledge."

Recent researches indicate that all matter as we know it has an electrical origin, explained Dr. Einstein. He said this has a remarkable bearing on the general theory of relativity and the physical properties of matter.

Professor E. P. Adams, of the physics department, in his resume of the lecture, said:

"There has, however, been a very serious difficulty involved in this view. For if we regard a portion of electricity which must be supposed to have a finite, although it may be a very small, volume, the repulsion between elements of that volume requires some unknown force in order that it may hold together."

Dr. Oscar Veblen, professor of mathematics, said: "Dr. Einstein does not insist he is right. He takes a detached view. He is just as interested in experiments that would disapprove his theory as in experiments that would prove it. He has the true scientific attitude."

The Evening Bulletin (Philadelphia), May 14, p. 21.

Note that Veblen's first name was Oswald, not Oscar.

14

Sees Boston

✦

Monday, May 9

Hebrew Societies Indorse Weizmann

Funds Urged to Provide a Jewish Homeland

At a conference in the West End Y.M.H.A. Hall, by 187 delegates, representing 28 cities, resolutions were adopted indorsing Dr. Chaim Weizmann, president of the World's Zionist Organization and the Keren Hayesod.

Benjamin Rabalsky, president of the New England Zionist Regional Union, presided and appointed a resolutions committee consisting of Harry Mannitz of Revere, chairman; Abraham Shocket of Dorchester, Jacob Promboim of Cambridge, Samuel Rugin of Malden and I. Segal of Brockton.

The chairman introduced Emanuel Newman, acting director of the Keren Hayesod Bureau of America, who said that it is the duty of every Zionist to stand firmly behind Dr. Weizmann.

Dr. Schmaru Levin of the World Zionist Commission spoke on the Jewish situation in Palestine. He recounted the heroism of the Jewish colonists who died defending their homes against the attacks of Arab marauders. He said the solution of the Jewish problem in Palestine was in a large emigration of Jews. "Let us raise the necessary money," said Dr. Levin, "and we can be certain of a Jewish homeland within the next few years."

Preparations are being made for the Weizmann-Einstein reception May 17 and 18. A banquet will be tendered to Prof. Weizmann the first day, and on the 18th there will be a demonstration in Mechanic's Hall, where Prof. Weizmann, Dr. Einstein, Dr. Schmaru Levin and Manacheim Mandel

Ussischkin will speak on the political, economic and cultural aspects of the new Jewish home in Palestine.

Jewish organizations of Metropolitan Boston have been invited to the reception of the Weizmann Commission. Hundreds of organizations will be represented at a conference in Lorimer Hall, Tremont Temple, tomorrow evening, where Dr. Schmaru Levin will speak.

The Boston Globe, May 9, p. 9.

Sunday, May 15

To Greet Einstein and Weizmann

More Than 5000 Pupils of Hebrew Schools to Join

More than 5000 pupils of Greater Boston Hebrew and Sunday schools and junior clubs will participate in the reception to the Weizmann-Einstein Commission Tuesday afternoon at 4, under the auspices of the Bureau of Jewish Education. The children will assemble at the corner of Elm Hill av and Seaver st, Roxbury, where the demonstration will take place.

Addresses in Hebrew and in English will be given by Rabbi H. H. Rubenovitz, Dr Leon S. Medalia, Louis Hurwitch and representative Boston Hebrew teachers. English and Hebrew songs will be sung by the children.

A feature will be the presentation of flowers by the children to Prof. Albert Einstein, Prof. Chaim Weizmann and their associates. Prof. Weizmann will reply in Hebrew. The official program is as follows:

TUESDAY

7 a.m.—Commission will be met at the South Station by a reception committee.

9 a.m.—Mayor Peters will give a breakfast to the visitors at the Copley-Plaza Hotel.

11 a.m.—Official greeting of the city on the steps of City Hall.

1 p.m.—Luncheon.

4 p.m.—Demonstration of 5000 pupils on corner of Seaver st and Elm Hill av, Roxbury.

5 p.m.—Mincha services at Adath Jeshurun Synagogue, Blue Hill av, Roxbury.

6:30 p.m.—A kosher banquet at the New American House. Addresses by Prof Weizmann, Prof Einstein, Menachem Mendel Ussischkin, Dr Schmaru Levin and Dr Ben-Zion Mossissohn.

WEDNESDAY

1 p.m.—Gov Cox will give a luncheon to the commission at the Hotel Touraine.

2 p.m.—Official greeting of the State from the Governor on the steps of the State House.

4 p.m.—Boston Zionist women headed by Mrs Isaac Harris, Mrs. Van Norden, Mrs. H. H. Rubenovitz, Mrs S. Myers will give a reception to Mrs Weizmann and Mrs Einstein. Place to be decided upon.

4 p.m.—American Academy of Arts and Sciences reception to Prof Einstein in Cambridge.

6:30 p.m.—New Century Club will give dinner to the visitors at the Boston City Club.

8 p.m.—Public reception and demonstration at the Boston Opera House.

8 p.m.—Public reception and demonstration at Mechanic's Building.

The Boston Globe, May 15, p. 13.

Monday, May 16

Kindred, Studying Relativity, Sees Some Light and Wants to Instruct Congress

WASHINGTON, May 17. The Einstein theory bobbed up in the House yesterday when Representative Kindred, New York, asked unanimous consent to extend his remarks in the Congressional Record on the "Non-Political Subject of Relativity," as advanced by the Swiss scientist.

Reserving the right to object, Representative Walsh, Republican, Massachusetts, declared that ordinarily matters in the Record were confined to things that one of average intelligence could understand. He asked Mr. Kindred if he expected to get the subject in such shape that the theory could be understood.

"I have been labouring earnestly with this theory for three weeks," replied the New York member, "and am beginning to see some light."

"What legislation will it bear upon?" Mr. Walsh inquired.

"It may bear upon the legislation of the future as to its general relations with the cosmos," said Sir Kindred.

Representative Longworth, Republican, Ohio, suggested that Mr Kindred ought to save discussion of the theory as applied to the relativity of political parties for a speech on the tariff.

The Boston Globe, May 17, p. 10.

Tuesday, May 17

Prof. Einstein to Arrive Here Today

Will Be Met This Morning by 200 Prominent Jews

Prof. Albert Einstein, world famous scientist, and Prof. Chaim Weizmann, head of the World Zionist Organization, will arrive in Boston at 7 o'clock this morning, and will be met by more than 200 prominent Jews in this city. The visitors will be accompanied by a delegation consisting of Manachem Mendel Ussischkin, Russian Zionist worker; Prof. Ben Zion Mossinsohn and Dr. Schmaryu Levin.

Mayor Peters will entertain the commission at breakfast at the Copley-Plaza, and at 11 o'clock the visitors will receive the official greeting of the city on the steps of City Hall. In the afternoon more than 5000 pupils of Greater Boston Hebrew schools will hold a reception at the corner of Elm Hill avenue and Seaver street, Roxbury, and will present flowers to members of the commission. Following the exercises the delegation will attend Mincha services at Adath Jeshurun Synagogue, Roxbury. There will be a banquet in the evening at the American House.

Harvard University, through President Lowell, will give the commission a reception at 10 o'clock tomorrow morning, and Gov. Cox will be the host at a luncheon at the Touraine at noon. In the afternoon Boston Zionist women will give a reception to Mrs. Weizmann and Mrs. Einstein at the Temple Mishkan Tefila, Roxbury. The celebration closes tomorrow evening with two mass meetings held simultaneously at the Boston Opera House and Mechanics building.

While in Boston the commission will appeal to Jews to support the Hebrew University to be erected on Mt. Scopus, Jerusalem, and for the Karem Hayesod, the Palestine foundation fund.

The Boston Herald, May 17, p. 2.

Boston Warmly Welcomes Profs Einstein and Weizmann

Mayor Peters and 100 prominent Boston men welcomed Prof Albert Einstein, the distinguished scientist, and Prof. Chaim Weizmann, head of the World Zionist organization, with members of their party, at the Copley-Plaza Hotel at 9:15 this morning.

Members of the Weizmann-Einstein Commission, who are touring the country in the interests of the Palestine Foundation Fund, arrived in Boston in two detachments.

Prof. and Mrs Weizmann, Dr. Ben Zion Mossinsohn and several others yesterday attended the convention of the Independent Order of Brith Abraham at Atlantic City. They reached New York too late last night to join the rest of the party, but they made Boston at 7 this morning, while Prof. Einstein's train was late. He did not get in until 8.

A committee of prominent Jewish citizens, 50 members of the Jewish Legion and a brass band were waiting for the visitors at the South Station. Many of the Jewish organizations were represented. The reception committee was headed by Isaac Harris, chairman, and A. L. Selig, vice chairman.

The more prominent members of the visiting commission are, besides Profs Einstein and Weizmann, Menachem M. Ussishkin, Russian Zionist worker, Dr Ben Zion Mossinsohn, head of the Hebrew High School in Jaffa, Palestine.

Profs Einstein and Weizmann Speak

The after-breakfast exercises were opened by the Mayor, who introduced the speakers.

Gov. Cox added to the welcome of the Mayor the welcome of the State.

Prof. Einstein was next called upon. Resounding applause greeted him. He made a short speech in German, telling of the great honor he felt had been bestowed upon the guests of the Jewish people by the wonderful reception. He further spoke of the work now being carried on in the institution in Palestine.

Prof. Weizmann was introduced by the Mayor, who lauded him for the invention of T.N.T. Prof. Weizmann made his address in splendid English. He said that Boston is a city where many of the people are learned and he praised the city for its wonderful institutions, saying that Boston is a spiritual and intellectual center.

[For details on Weizmann's address, see pp. 220-221.]

Mayor Peters's Address

The welcoming speech of Mayor Peters follows:

"It is my great privilege to voice the welcome of this city to Prof. Einstein and Prof. Weizmann. Even if our distinguished visitors were here merely upon private business, we should be proud to do them honor, both for their personal distinction and as members of a race so eminent in every field of creative scholarship.

"But they come to us in a representative character, their primary purpose being to collect funds for a Jewish university in Palestine. This is a mission with which the people of Boston can readily sympathize, for we are the descendants of colonists, who, within six years of their settlement here, had founded Harvard College, now the foremost of American universities. We may find a happy parallel in the circumstances that the Jewish Commonwealth, like our own, is to lay down and establish, as one of its foundation stones, an institution dedicated to the preservation and the extension of knowledge.

"In a wider sense our visitors come to promote the cause of Zionism. One does not need to be of Jewish blood to appreciate this movement. For 2000 years the Jewish exiles have turned their faces toward Jerusalem, lamented their dispersal and dreamed of a reunion in the land of their fathers. Today we see the Turk never acclimated there, always an alien, taking his departure and the children of the soil returning to their own. It is one of the most dramatic episodes in all history. One of the great tragic stories of the world approaches what we hope will prove to be its happy consummation.

Land Flowing with Milk and Honey

"It is the same land that the returning exiles will behold. The valleys of Sharon and Jericho are as fertile now as when they flowed milk and honey in biblical times. The vine, the olive and the fig-tree only await the coming of skillful husbandman to bear fruit in their old abundances. The cedars of Lebanon may be shorn away but the plains below stand ready again for the

Jewish shepherd and his flocks. In reclaiming this country from the neglect of ages and in building there a durable State, the Jew has an opportunity at last for a collective expression of the genius of his race, which has shone so brilliantly in individual examples.

"Who can doubt that with his intelligence, his enterprise, his industry, his indomitable force, he will overcome whatever obstacle may lie in his path? If difficulties could daunt him, his race would have been exterminated or absorbed centuries ago; but never was it more potent, more in the front of human endeavor, than it is at this hour.

"For those of us who are Christians, there is a gracious side to the movement, in that the reentry to Palestine will come, not by conquest but by general consent. It is in a way an act of reparation, by the Nations of Europe for the injuries they have inflicted in the past. We may fairly regard it as an acknowledgment of grave injustice, which offers the hope, at least, of a great reconciliation.

Debt World Owes to Israelites

"Viewed in that light, the new Zion will be something more than a refuge for persecuted Jews or even a symbol of Jewish unity. It will express also the incalculable debt the world owes to the children of Israel. From the days of Elijah and Isaiah their moral fervor blazed out like an inextinguishable fire. Their penetrating intellects have led the way in more than one branch of scientific discovery.

"It is this conjunction of qualities that has given the race its unique and precious character. Its thinkers may dwell on the solitary heights, but its prophets belong to all humanity. Not many of us can follow Prof. Einstein in his discussion of the mathematical properties of space; but all of us can understand his refusal to sign the manifesto of the 93 German professors, disclaiming responsibilities for the war. The clear sense of justice which prompted that act, the fine integrity and independence behind it, were essential parts of his Jewish heritage. They are reasons no less than his intellectual eminence, why we honor him and his cause today.

"The world has need of men of his mold—such tolerant and impartial as between factions, but burning with a passion for righteousness. This is the type that has made Israel great. It is the type of teachers that must arise if we are to solve our social problems."

Men Present at the Breakfast

Among those present at the breakfast which Mayor Peters gave were the following: Abraham Alpert, Louis Baer, Hon Edmund Billings, Robert J. Bottomly, David J. Brickley, Alexander Brin, Edward J. Bromberg, L. B. Buchanan, J. Paul Canty, George Canty, Runert S. Carven, John M. Casey, T. J. Cassidy, J. Cohen, Edwin F. Collins, Walter L. Coliths, Sidney H. Conrad. Hon Channing H. Cox, George J. Cronin, Frank S. Deland, Philip S. Davis, Walter W. Dickson, John A. Donoghue, James Donovan, Carl Dreyfus, Julius Eisemann, David A. Ellis, Rabbi L. M. Epstein, Francis J. W. Ford, Benjamin Freedman, A. H. French, Irvin McD. Garfield, Rabbi H. R. Gold, Henry E. Hagan, I. Hamlin, Isaac Harris, Hon. Robert O. Harris, W. C. S. Healey, Simon E. Hecht, Arthur D. Hill, Forrest P. Hill [?], Albert Hurwitz, Charles H. Innes, [Carl] Kaffenburg, Edward T. Kelly, F. J. Kneeland, John L. Kelly, Abram Koshland, Daniel W. Lane, Henry H. Levenson, Joseph F. Lompey, R. C. McCabe, Edward F. McLaughlin, Joseph P. Manning, Hon. Nathan Matthews, John A. Meade, Dr Julius Meyer, James T. Moriarty, Charles A. Morse, L. H. Murlin, Joan R. Murphy, Malcolm E. Nichols, George R. Nutter, Thomas C. O'Brien, E. V. B. Parke, James J. Phelan, Arthur S. Pier, Jacob Promboim, Benjamin Rubalsky, A. C. Ratshesky, James W. Reardon, Dr. Milton Rosenau, Rabbi H. H. Rubenovitz, Samuel Rubin, Ellery Sedgwick, Harris Selig, James D. Shea, Hyman Shocket, Rabbi David M. Shehet, Robert Silverman, Samuel Silverman, M. J. Slonim, Thomas F. Sullivan, Frank Supple, Elihu Thomson, Felix Vorenberg, Simon Vorenberg, Bentley W. Warren, Jacob Wasserman, James A. Watson, Edwin S. Webster, Arthur H. Wood, C. F. Weed, Walter White, James T. Williams Jr, Herbert A. Wilson, Jacob L. Wiseman.

Seated at the head table were Robert Silverman, Charles A. Morse, Prof M. M. Ussischkin, George R. Nutter, Dr Schmaryu Levin, Edmund Billings, Prof Einstein, Mayor Peters, Gov Cox, Prof Weizmann, Nathan Matthews, Prof Elihu Thomson, U.S. Dist. Atty Harris, Dr Ben Zion Mossinsohn, Dr L. H. Murlin and Isaac Harris.

Another Formal Welcome at City Hall

Mayor Peters extended another formal greeting to the Einstein-Weizmann party on the City Hall steps when it came down about 11 a.m. Full, 500 men and women crowded the yard and there were enthusiastic cheers for the party as it came in view. Mr. Peters here welcomed the party as of the group "which is to carry out the noble aspiration of a race to have its own free country."

Responding for the party Mr Weizmann described it as "happy, moved and proud to receive such a handsome welcome in Boston. Your city is indeed a great seat of liberty," he said. "We shall learn from you how to build up a glorious, free community."

After posing for the photographers, the party entered waiting automobiles, to continue the day's itinerary.

Program of Entertainments

Boston Jewry will take the leading part in entertaining the visitors during their two-day stop in this city. The program is as follows:

4 PM—Demonstration by 5000 pupils of Bureau of Jewish Education's Hebrew and Sunday schools, corner of Seaver st and Elm Hill av, Roxbury.

5 PM—Mincha services at Adath Jeshurun Synagogue, Blue Hill av.

6:30 PM—Kosher banquet at American House, with addresses by Profs Einstein and Weizmann, Menachem Mendel Ussischkin, Dr Schmaryu Levin and Dr Ben-Zion Mossinsohn.

WEDNESDAY

10 AM—Reception by Pres Lowell at Harvard University.

1 PM—Luncheon tendered the visitors by Gov Cox at Hotel Touraine, followed by official greeting of State at State House.

4 PM—Reception to Mrs Einstein, Mrs Weizmann and ladies by Boston Zionist Women at Mishkan Tefila Temple.

4 PM—American Academy of Arts and Sciences reception to Prof Einstein at Cambridge.

6:30 PM—Dinner given by New Century Club at Boston City Club.

8 PM—Public reception and demonstration at Boston Opera House.

8 PM—Public reception and demonstration at Mechanic's Building. Members of the commission will alternate in addressing both meetings.

The Boston Globe, May 17, pp. 1, 10.

Mayor Andrew James Peters must have enjoyed the opportunity of a warm get-together, as he took to such occasions. "A Brookline squire endowed with three apostolic names," as his predecessor James Curley characterized him, he was an amiable, high-principled person, a sailor and horseman, always tanned.

Among the personalities sitting at the head table, Elihu Thomson, president of MIT and one of the founders of General Electric Company, and Lemuel Herbert Murlin, president of Boston University, deserve mention.

Mrs Weizmann Likes Her Cigarette

This is to let all the world know, including Boston girls who slyly like their cigarettes, that Mrs Olga Weizmann, wife of the visiting scientist who invented tri-nitro-toluol (TNT), likes her after-breakfast smoke.

Public Works Commissioner Tom Sullivan anticipated her wish for one at the Mayor's Copley-Plaza breakfast to the visitors this morning, and scored heavily in a social way with Mrs Weizmann and six or eight other men seated about the breakfast board.

Through the courses up to cafe noir, Mrs Weizmann had been entertaining the company with the engaging story of her life and meeting with the scientist—her birth in Russia, her education in Switzerland and England, and her winning of degrees in medicine and surgery.

On came the cafe noir. Everybody smacked his or her lips in gustatory satisfaction over it, and naturally, all the men hankered for the usual smoke.

When the conversation had become a bit strained for the lack of nicotine, Maj Sullivan realized it. The "Major" is no globe trotter, but instinctively, he knows how to do things handsomely, in approved "Continental" style. He extended his gold cigarette case to Mrs Weizmann, who gratefully picked out one.

Then he offered a match. Then a happy renewal of talk all around, among those named and Councilors Ford and Collins, Purchasing Agent Cronin, C. F. Weed and Institutions Commissioner O'Brien, who were of the party.

The Boston Globe, May 17, p. 1.

Olga is a nice name for a Russian woman, but Mrs. Weizmann's first name was Vera. According to the following report, she was offered a cigar that she refused and took her own cigarette. A cigar for a lady? I prefer the first version of the story.

Einstein Arrives

Scientist Gives a Demonstration in Relativity—Correlation of Time and Place for Breakfast—Governor and Mayor Extend Cordial Welcome—Bostonians Greet Party at Copley Plaza

Einstein's theory of relativity was applied to the breakfast which the city of Boston gave at the Copley-Plaza Hotel this morning in honor of the noted scientist and his associates. The breakfast was there at the specified hour, 9 o'clock, and Mayor Peters and about seventy-five guests were present, but this hour had been fixed by the ancient rule of time which governs Boston activities and had only a relative influence on the movements of the party in which Professor Einstein is the centre of attraction. Time and place as readjusted by Einstein in their relation to each other coincided at a point which placed the Einstein group in the Copley-Plaza Hotel at 9.30 o'clock, and a few minutes later there was a perfect conjunction of guests of honor, other guests and breakfast.

Relativity had been carried to the extent of overcoming the forty-five minutes which Professor Weizmann had lost on the way to Boston, reducing those forty-five minutes to thirty minutes before Einstein and Weizmann appeared at the Copley-Plaza Hotel together, and the cordiality of the reception and brevity of the speeches negotiated time so smoothly that by old-fashioned Bostonian calculation the party might have overcome losses and have been at City Hall on schedule time for the official celebration of the day.

It was really at the South Station that the visiting scientists had their first introduction to the Boston public, for there was a throng of people on hand to receive them, and Einstein had forty-five minutes to wait for Weizman to arrive; but the first official reception was given at the Copley-Plaza.

Mayor Peters Presides

Accompanying Professor Einstein to the breakfast were his wife, Dr. Chaim Weizman and his wife, and several others who have come from Palestine on a mission in connection with the proposed Hebrew University on the Mount of Olives, in Palestine.

Mayor Peters presided, and seated with him at the head table were Governor Cox, Professor Einstein, Edmund Billings, Dr. Schmaryin Levin, George R. Nutter, Professor M. M. Ussischkin, Charles A. Morse, Robert

Silverman, Dr. Weizmann, former Mayor Nathan Matthews, Professor Elihu Thomson, Judge Robert O. Harris, Dr. Ben-Zion Mossinsohn, Dr. L. H. Murlin and Isaac Harris.

Mrs. Einstein and Mrs. Weizmann, both of whom speak English fluently, were seated at a table in front of the head table, conversing with a number of Boston business men and lawyers, and when the cigars were passed around Mrs. Weizmann politely declined and lit a cigarette instead.

Grace was said first in English and then in Hebrew by Rabbi H. H. Rubenovitz of the Temple Mishkan Tefila of Boston. Four brief speeches were made. Mayor Peters, as presiding officer, dwelt somewhat upon the significance of the movement to establish a Hebrew University in Palestine, in the interest of which the commission has come to Boston, but waived the privilege of discussing the Einstein theory; Governor Cox welcomed the visitors to the State and complimented the Jewish race upon its contribution to the industry, commerce and spirituality of the Commonwealth. Professor Einstein, speaking in his native tongue, merely emphasized the importance of the intellectual movement for Palestine and thanked the mayor and the governor for the reception. Professor Weizmann spoke in English, referring to Boston's intellectuality and expressed a hope that he may be permitted to welcome some of the distinguished Bostonians at the proposed university on the Mount of Olives in the course of perhaps five years.

[Here follows a short version of Mayor Peter's address. The full text is presented on pp. 214–215.]

Dr. Weizmann Responds

Dr. Chaim Weizmann, who was introduced as the inventor of TNT, spoke in English, with clear and almost correct enunciation, saying in part:

"We are deeply impressed by your cordial welcome, Your Honor. We were indeed anxious to come to Boston, particularly because we Jews feel that we can learn much from Boston, and that is my special compliment to Boston, which I will explain.

"For we Jews have been through many schools during our long orphanage. We have seen and learned a great deal during the many centuries of our wanderings, but we recognize Boston as the power station radiating special influence all over the world because it has performed the peculiar task of transplanting an ancient civilization into a new setting. It is this performance which makes Boston the seat of culture in the sense in which we understand it in Europe. This is why Boston can't be measured simply by square kilometers, but rather by its layers of civilization. In this respect we who are devoted to Palestine desire to perform a similar task, to wit, to transplant an ancient civilization which we have been carrying with us for many centuries, into a new-old country.

"If Palestine were measured merely by the square kilometers, it is indeed a puny little country, but if it is to be measured by the centuries of history, and its layers of civilization, it is a mighty kingdom. Our purpose, therefore, is to accomplish in Palestine what you have already accomplished in Boston, namely, transplanting a high degree of civilization and thereby making it an international seat of culture.

"We thank you, therefore, not only for the tremendous financial and moral support which you have given us at all times, particularly in times of stress, but for your present kindly welcome and sympathy and interest in our great undertaking.

"And if I may express a wish, at present only a dream, it is this: That the time may come when we shall be able to receive the mayor of Boston, the governor of Massachusetts and the distinguished professors and merchants and citizens of this city and State, some bright day on the top of the Mount of Olives, which we are now climbing with so much difficulty. It is a hard road to travel, but when we reach the top we hope to have the opportunity of reciprocating in a measure in Palestine the hospitality which you have so generously accorded us this morning."

The Boston Evening Transcript, May 17, pp. 1, 7.

Just What It Is

Einstein Shows How Both Time and Space Would Vanish Under Certain Hypotheses within His Relativity Theory—Very Simple

Who should sit next to Professor Einstein at the breakfast was a matter of concern, for it was generally known that he cannot speak English and equally well known that nobody understood his "relativity" theory which has made the man famous. It finally fell to the lot of former Mayor Nathan Matthews to occupy the seat, after he had protested with Mayor Peters, and it was discovered that Mr. Matthews and Professor Einstein had found both a common language and a subject to discuss.

On many occasions Mrs. Einstein came to her husband's assistance as an interpreter as he was engaged in conversation by other guests after the breakfast. Whether time be real or merely a relative conception it was used by the visitors to good advantage during the remainder of the day, first for a visit to City Hall for another official welcome by the mayor of the city, then for an automobile drive through the park system and a little later in an earnest effort to explain "relativity" to a group of newspaper men at Hotel Touraine.

To state in ordinary language, what "relativity" is might confuse the scientists and Professor Einstein himself balked at the undertaking to illustrate the idea by any reference to concrete objects. When that was suggested to him he stared into space and shook his head, as if there were nothing in space to aid him.

"The basic principle, or new idea, is that there are no privileged states of motion in nature, and the most interesting consequence is that time and space which formerly were considered as prioric concepts, cannot now be considered as separate from physical events."

"If matter were destroyed, then, what would happen to time and space?"

Quickly the answer came: "Then there would be no time and no space."

In this connection, as the professor discussed it, time is as real as length, and the inference may be drawn that if there is nothing to measure there is no length, which aids the conception of the disappearance of both time and space if matter vanishes.

Of course that seemed very plain and the interviewers became apprehensive of its effect upon the multitude and they welcomed the assurance that:

"It will have no effect upon our everyday life at present."

Yet this theory could not be dismissed so readily, for Professor Einstein admitted that he is giving his life to its development, realizing that he has not yet sounded its depth, and he explained that it will affect research work in the future, and will have a large effect upon our understanding of astronomy, especially the fixed stars.

On his present visit Professor Einstein may remain in Boston until next Friday. It is partly his object to arouse interest among the Jews in America for the Hebrew University which is to be established in Jerusalem, and the first undertaking is to "create a mood, or state of mind as a basis for the collection of money later." An organization has already been started of Jewish

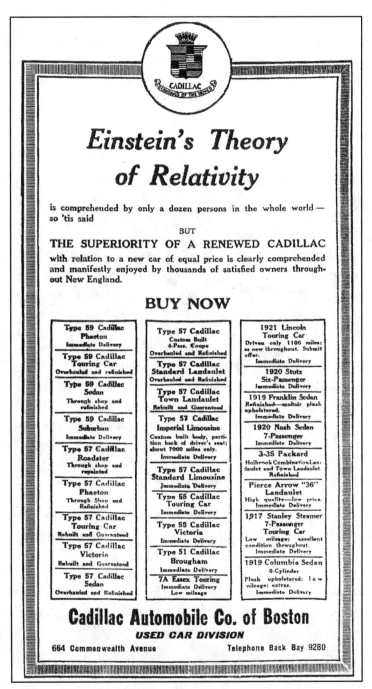

physicians who are interested in the project and will raise $1,000,000 which is regarded as enough for a beginning. This organization will make its first public announcement at a banquet in New York on May 21.

Professor Einstein will be associated with the new university in some way, but he does not purpose to make it the centre for the promotion of his special theory.

The Boston Evening Transcript, May 17, pp. 1, 7.

Einstein Sees Boston; Fails on Edison Test

Asked to Tell Speed of Sound He Refers Questioner to Text Books

Special to *The New York Times.*

BOSTON May 17.—There was a large crowd at the South Station this morning to greet Professor Einstein of relativity fame and his party. From the station the visitors made an unexpected automobile tour through the north and west ends, Boston's Jewish quarters and then proceeded to the Copley Plaza Hotel, where they sat down to breakfast with Governor Cox, Mayor Peters and some 75 distinguished guests.

Mrs. Weisemann, wife of Dr. Chaim Weisemann, of the visiting party, surprised the party when it came time to pass around the cigars by calmly producing a cigarette and lighting it. Her action was welcomed by the men. They wanted to smoke but hesitated to do so in the presence of Mrs. Weisemann, and Mrs. Einstein, the only women present. Mrs. Weisemann's action in "lighting up" paved the way and the men lit their cigars.

Professor Einstein gave out through his secretary the following message for Bostonians:

"I am happy to be in Boston. I have heard of Boston as one of the most famous cities of the world and the centre of education. I am happy to be here and expect to enjoy my visit to this city and Harvard."

Of course the famous visitor had to run into the ever-present Edison questionnaire controversy. He did not tackle the whole proposition but so far as he went failed and thereby became one of us. He was asked through his secretary, "What is the speed of sound?" He could not say off-hand, he replied. He did not carry such information in his mind but it was readily available in text books.

Professor Einstein took issue with the famous inventor's contention that a college education is of little value. Professor Einstein said he believed education was a good thing. If a man had ability, he thought, a college education helped him to develop it. He stated he had not had an opportunity to study the Edison list of questions. He had heard of the American inventor in connection with the invention of the phonograph and electrical appliances.

Mrs. Einstein said that while Edison was an inventor who dealt with practical and material things, her husband was a theorist who dealt with problems of space and of the universe.

The New York Times, May 18, p. 18.

Mrs. Einstein's remark at the end might have sounded sharper if she had known that at the time Edison was not impressed by her husband's achievements. When in 1923 his son, Theodore, showed more talent for mathematics and physics at MIT than for business management, Edison remarked: "Theodore is a good boy, but his forte is mathematics. I am afraid . . . he may go flying off into the clouds with the fellow Einstein. And if he does . . . I'm afraid he won't work with me."[67]

Edison's lifelong doubts about the value of college education were reinforced by the fact that, although he expected job candidates to give correct answers to at least 90 percent of the questions (!), only 60 out of 2000 college-educated people were up to scratch. "There is something wrong with the college system. The present system does not train men to think," he remarked. The candidates, however, must have been happy that Edison had not included questions about the material Gutenberg's letters were made of (Edison thought it was wood, and not brass) or about the nationality of Richard Wagner, to which he replied: "He was a Jew," an answer which, had it been correct, would have deprived Hitler of his favorite composer.[68]

But what about Edison's questionnaire?

To test the knowledge of job applicants, Edison had made questionnaires for each position, consisting of 150 items, with 2.5 hours to answer. A college graduate later listed 140 questions from memory. Here are a few samples: "What country consumed the most tea before the war?" "With what metal is platinum associated when found?" "In what cities are hats and shoes made?" "Who invented the printing press?" "How is leather tanned?" "Of what kind of wood are kerosene barrels made?" "What part of Germany do we get toys from?" And, of course, the question Einstein was asked: "How fast does sound travel per foot per second?"[69]

Just a few days earlier, a reporter of the *Boston Herald* put twenty of Edison's questions to six leading personalities, among them to Mayor Peters and Governor Cox. Here are the questions and Governor Cox's answers as published in the *Sunday Herald* of May 15:

1. Where do we get shellac from? From a can.
2. What is a monsoon? A funny-sounding word.
3. Where do we get prunes? Breakfast.
4. Where do we get domestic sardines? Wellfleet.
5. Where do we import cork from? Canada.
6. Of what kind of wood are axe handles made? Hardwood.
7. Who wrote "Home, Sweet Home"? A happy man.
8. Where are condors to be found? In none of our state reservations.
9. Who was Cleopatra? A woman of temperament.
10. What voltage is used in street cars? A diminishing voltage.
11. What is felt? Something not straw.
12. What states produce phosphates? [Answer:] ??
13. Why is cast iron called pig iron? This is No. 13—so I shan't answer it.
14. Where is Spitzbergen? Hidden by snow.
15. Who was Francis Marion? I don't know.
16. Who invented logarithm? A tyrannical Dartmouth professor.
17. What ingredients are in the best white paint? Soot and dirt.
18. Where do we get benzol from? Benzuela.
19. How is window glass made? So that it is transparent.
20. Bound the state of West Virginia. On the north by $16 a ton; on the east by $16 a ton; on the south by $16.50 a ton and on the west by $20 a ton.

Einstein Party Rallies Jews to Aid of Zion

Visit of Noted Leaders Brings Many Pledges of Support for Project—Scientist Gives Outline of Theory

Dr. Albert Einstein, famous for his theory of "relativity," and Dr. Chaim Weitzman, inventor of deadly TNT, were the guests and to all intents and purposes the idols yesterday of the Jewish population of Boston. They arrived in the early morning and all day were followed about the city by hosts of their people, who feted and feasted them, and with whom they joined in worship and in Zionist demonstrations.

They came, not talking about new laws of physics which take account of infinity, nor of baleful high explosives, but of the nationalistic and spiritual hopes of their race. Dr. Einstein, in particular, brought as his message the aspiration that Palestine shall become a world renowned centre of learning, not alone for Jewry but for the entire world.

500 Attend Banquet

These thoughts gave the keynote of the gathering of upwards of 500 Jewish people, men and women, who attended the banquet given in honor of the visitors at the American House last evening. "Self-taxation" was the new law that Dr. Weitzman proclaimed, by means of which Jewry of America must redeem its brethren in other lands from their bondages and set up a home and a fane for them in the Holy Land, the ancient home of their kind.

Then and there the "self-taxation" was begun and so generous was the response of the diners, representing many Jewish organizations and congregations both in and outside Boston, that hundreds of thousands of dollars are expected to accrue to the Keren Hayesod, the Palestine foundation fund, as a result of the visit of the Zionist leaders to Boston.

"The moment has arrived," Dr. Weitzman declared to the diners, "for the concentration of Jewish effort upon the upbuilding of the National Jewish Home—Palestine. We begin our work at a great and tragic hour. While our opportunities in Palestine are enheartening, two-thirds of the Jewish race in Eastern Europe are living at this moment under intolerable conditions. The Jews of America are providentially the remnant that may now liberate the larger part of Israel."

The direct appeal for funds adequate to realize the Zionists' ideal of a re-constructed Palestine, was conducted by Rabbi H. Raphael Gold of Temple Adath Jeshurun.

"With the wholesale slaughter of Jews in Europe, and with the doors of this country closed to them, their only hope," he said, "is in Palestine."

Dr. Einstein spoke feelingly in German in behalf of his ideal, his dream, of making Palestine a lamp of learning, a centre of the intellectual life, not merely of Israel, but of Christendom, of civilization. The new Jewish university built upon the Mount of Olives, outside Jerusalem, already do-ing notable work in teaching Jewish youth mechanic arts and handicrafts, the products of which are making a showing in the world markets, is his pet project, and already, as the result of the visit of the Weitzman-Einstein commission to this country, the Jewish medical men in America have un-dertaken raising $1,000,000 for the institution.

Thousands at Station

Isaac Harris was toastmaster at the banquet. Dr. and Mrs. Einstein ac-companied by Dr. Ben-Zion Mossinsohn, head of the Hebrew high school at Jaffa, Palestine, M. M. Ussischkin of the world Zionist commission and Mrs. Weitzman, arrived at the South station shortly after 7 A.M. yester-day and were given a tumultuous greeting by thousands of Jewish people. A detail of police intended to keep the crowd from overwhelming them was late through misunderstanding, but some 100 former members of the Jewish legion, Jewish soldiers in the British service, acted as an escort. Dr. Weitzman arrived about 45 minutes later, having been delayed by a visit he had made to Atlantic City. The party was taken in automobiles through the Jewish districts of the North and West ends, being given enthusiastic greet-ings along the route. They were taken to the Hotel Touraine, and registered, a crowd following them into the lobby.

At 9 A.M. Mayor Peters was their host at breakfast at the Copley-Plaza. Later he greeted them officially on the steps of the City Hall. At 4 P.M. the men of the party were escorted by leading Jewish citizens, in automobiles to the State House, a guard of 20 of the Jewish Legionnaires attending them under Sergt. Maj. "Jack" Williams, a military medal veteran of the British ser-vice. They were received by Gov. Cox, who insisted that the two distinguished Jewish scholars sit down and rest while chatting with him, as he said that he well understood the weariness resulting from their constant traveling.

Visit Roxbury

After a pleasant quarter-hour they were driven to the site of the proposed new Temple Mishkan Tefila, in Roxbury. Dr. H. H. Rubenovitz, rabbi of Temple Mishkan Tefila, greeted them and addresses were made in English by Dr. Leon A. Medalia, president of the bureau of Jewish religious schools, and in Hebrew by Louis Horwitch, superintendent of this bureau, and by Dr. Mossinsohn. The presence of upwards of 5000 children, pupils of the schools conducted under the auspices of this bureau, who sang national and devotional songs in Hebrew, gave the gathering a festal air, which was enhanced by the display of innumerable small American and Jewish republic flags. The latter are white with two horizontal blue stripes, and in the middle, in blue, the six-pointed star, the symbol of the Jewish people and faith.

Dr. Weitzman, in his own behalf and for his colleagues of the commission, was presented a huge floral piece in blue and white, the same six-pointed star being the central device. It was presented by two little girls—Edith Bellen of the Grace Aguilar school and Janice Levy of the Herzl school.

At 5 P.M. "mincha" services were held at the Temple Adath Jeshurun Synagogue in Blue Hill avenue. Prayer was chanted by Cantor Abraham Brodski, and Rabbi Gold called upon the Jewish people to be ready at all times for service. Ezra Goldberg, vice president of the congregation, delivered a short address, at the close of which he announced that the congregation had donated $1000 to the Zion Foundation fund.

Many Engagements Today

The ceremonies in honor of the noted Hebrew scholars will be continued today, beginning with a reception to them by President Lowell at Harvard and a reception at the Harvard Union. Gov. Cox will be their host at luncheon this noon at the Touraine. At 4 P.M. Boston Zionist Women will give a reception to Mrs. Weitzman and Mrs. Einstein, at Temple Mishkan Tefila, Roxbury. At 4 P.M. the American Academy of Arts and Sciences will give a reception to Dr. Einstein in Cambridge, and at 6:30 P.M. the New Century Club will give the visitors a banquet at the City Club. At 8 P.M. there will be public demonstrations at the Boston Opera House and at Mechanics building, the visitors speaking successively in the two meetings.

Dr. Einstein's visit occasioned a great deal of interest regarding his little understood theory of relativity. When seen in his rooms in the Touraine he was reluctant—almost diffident—about speaking of this theory, saying that if he were able to express himself satisfactorily and if his words—he speaks in German—were translated with scientific accuracy, the general public even then would be unable to comprehend it fully. Finally he was persuaded to say what, in a general way, the new theory means, and he replied, through an interpreter:

"It shows that time and space have no independent existence from matter."

"What will be the ultimate scientific outcome of that?" Dr. Einstein was asked:

"The main thought is that there are new ideas about time, space and gravitation," he replied.

"Are there many scientists in the world who disagree with your theory?" was the next question.

"As far as I know," came the response, "the best known theorists are believers with me."

The Boston Herald, May 18, pp. 1, 8.

Einstein Tells Why He Can't Explain

Only Dozen People in World Understand Idea Anyway—Zionist Party Feted All Day and Evening by Boston Admirers

If you have read everything that has ever been written on the Einstein theory—even Prof Einstein's own book on that subject—and if you have talked with everybody who professes to know all about it—even Prof Einstein himself—and you find that you are still in the dark as to just what the famous German theorist's theory is, don't feel disturbed or that your lack of understanding is any reflection on your intellectual capacity. It isn't—necessarily—and there are millions of people who are out in the darkness with you.

Prof Einstein himself has said that there aren't 12 men in the world who thoroughly understand his theory.

"The theory is relativity."

That is about as far as the average person cares to go offhand.

Various theorists, philosophers and scientists in all parts of the world have undertaken the task of explaining to a curious and eager world just what it is all about. And some of their explanations of the Einstein theory have been wonderful to contemplate. One eminent authority recently made the whole thing perfectly clear with the following explanation:

"The time-space concept clarifies our idea of time, which may be defined as a fiction of the finite and three-dimensionally-conditioned intellect devised to differentiate, locate and dislocate events in the cosmic change and experience.

"Time is linear and directed; it is therefore reentrant, an infinity of time being circular and returning upon itself. I infer that hyper-dimensional time possesses qualities of superficiality such as spreading or stretching; a hyper-dimensional (v or 'N') time-infinity is spherical in the three-space and dislocated events or changes in three-space are hyper-dimensionally synchronous."

Wishing to get something even clearer than this, the *Globe* yesterday decided to ask Prof Einstein himself for an explanation of this theory and before dinner last evening a *Globe* reporter had a brief interview with the sponsor of the relativity theory.

Courtesy Personified

Prof Einstein is courtesy personified. He gives the impression of being rather small in stature, though really he is of medium height, and inclines toward stoutness. His long, bushy hair, worn a la the William Jennings Bryan of other days, is streaked with gray. His eyes are brown and bright. He has a mustache.

Though his theory of relativity pertains to many things, such as light, space, gravitation, etc, this eminent German has found one thing since he came to America about which there is nothing relative. That is weariness.

Prof Einstein is pretty nearly "all in," and his wife, who accompanies him on his tour, was confined to her bed at the Hotel Touraine yesterday afternoon.

Prof Einstein speaks no English, so all conversation with him must be carried out through an interpreter, unless one happens to speak German fluently.

Perhaps it was the reporter's imagination, but he thought he discerned a pitying look in the eyes of the learned German when he was informed that the newspaperman wished to talk to him about the Einstein theory. The reporter certainly felt sorry for himself.

Throughout yesterday all the prominent men Prof Einstein had met had very carefully avoided the theory of relativity in their conversation with its author. The Mayor "ducked" it. So did the Governor. So did almost everybody else.

Untranslatable

Very timidly the reporter asked Prof Einstein if he would please explain to the *Globe* readers just what his theory is.

The professor talked for a moment in German to Mrs Munsterberg, who very kindly volunteered to act as interpreter. Then she began to explain:

"He says it is too fine, the shades of meaning are too delicate to be translated.—"

Here the professor interrupted her, and after listening for a moment, Mrs Munsterberg began again:

"He asks me to explain," she said, "that if he could tell it absolutely correctly and if an absolutely correct literal translation could be given, even then the public would not understand. Each language has its own peculiarities, and they cannot be put into another language."

Here Prof Einstein spoke again, and as he concluded Mrs. Munsterberg went on:

"His theory is that time and space and gravitation have no independent existence from matter."

"But what will this theory ultimately work out to; or, to be quite blunt, what will it accomplish for the world?" asked the interviewer, completely dizzy and far beyond his depth.

When this question was translated to him, Prof Einstein smiled (and it look like a pitying smile, too), and he talked again at some length.

"The main point of it all is that these are new ideas about time, space and gravitation"—was the reply.

"O," said the interviewer.

Philosophers Agree, Anyway

Then, out of the goodness of his heart, Prof Einstein declared that of all the reports that have been printed regarding his renowned theory by papers in all parts of the world, not one of them has ever clearly explained the theory so that it could be fully understood by the general public.

"Have you found many scientists and philosophers who disagree with your theory?" he was asked.

"On the contrary," he replied, "the best-known theorists in the world are firm believers with me."

Then there was some more discussion, until it was obvious that Prof Einstein was weary of trying to explain. What he said comes down to this:

"According to the theory of relativity, there are no privileged states of motion in nature, and space and time, which we once considered independent, now are considered inseparable from the rest of matter. Which is to say that if all matter were destroyed, there would be neither time nor space."

If that is the correct explanation of the Einstein theory, then the *Globe* is the first paper in the world to correctly explain it to the public.

The Boston Daily Globe, May 18, pp. 1, 6.

> Sorry, *Globe*, the *Boston Evening Transcript* had already published this sensational explanation the day before (see pp. 221–224).
>
> As for Einstein's hair-mate, William Jennings Bryan, he was a three-time Democratic presidential candidate around the turn of the twentieth century and, as secretary of state under Woodrow Wilson from 1912 to 1915, a strict neutralist in World War I.

Boston Welcomes Zionists, Monster Banquet at Night

Boston yesterday extended a royal welcome to two distinguished scientists, Prof Albert N. Einstein, originator of the famous Einstein theory, and Dr Chaim Weizmann, inventor of TNT, who is also head of the World Zionist organization. It is in behalf of the Palestine Foundation Fund that the two famous men, together with their wives, and a sizable party working for the same cause, are in this city.

The visitors were greeted yesterday morning on their arrival by a large gathering, and from the South Station they proceeded at once to the Copley-Plaza Hotel, where they were the guests at breakfast of Mayor Peters and 100 prominent business men.

From the moment of their arrival in this city they were kept busy attending functions of a varied nature. Following the breakfast they visited City

Hall and later the State House. At 4 p.m. they attended a demonstration by 5000 pupils of the Bureau of Jewish Education's Hebrew and Sunday schools at Elm Hill av Roxbury. Then they attended the Mincha service at Adath Jeshurun Synagogue, Blue Hill av, at 5 o'clock, and at 6:30 were the guests of honor at a Kosher banquet given at the American House.

The members of the Weizmann-Einstein Commission, as it is called, arrived in Boston in two detachments yesterday morning. Prof and Mrs Weizmann, Dr Ben Zion Mossinsohn and other members of the party had on Monday attended the convention of the Independent Order of Brith Abraham in Atlantic City, and though they were too late in reaching New York Monday night to catch the train taken by Prof Einstein and his party, they reached Boston at 7 yesterday morning, while Prof Einstein and his party did not get in until 8.

They were greeted at the South Station by a committee of prominent Jewish citizens, 50 members of the Jewish Legion and a brass band. The reception committee was headed by Isaac Harris and A. L. Selig.

The more prominent members of the visiting commission are, besides Profs Einstein and Weizmann, Menachem M. Ussishkin, Russian Zionist worker, Dr Ben Zion Mossinsohn, head of the Hebrew High School in Jaffa, Palestine.

Welcomed at Breakfast

Mayor Peters presided at the breakfast at the Copley Plaza and introduced the speakers. Gov Cox added the welcome of the State to that of the city. Prof Einstein made a short speech in German.

Prof Weizmann made his address in splendid English, saying that Boston is a city of learning and praising the city for its wonderful institutions. He said, that Boston is a spiritual as well as an intellectual center.

The Mayor praised the Jewish people and in the course of his address, in touching upon Prof Einstein, he said: "Not many of us can follow Prof Einstein in his discussion of the mathematical properties of space but all of us can understand his refusal to sign the manifesto of the 93 German professors, disclaiming responsibility for the war."

Later the Mayor extended another formal welcome to the visitors when they visited City Hall about 11 a.m.

In the afternoon, the party visited the State House, when they were cordially greeted by the Governor. The party was escorted by a group of Jewish-American boys who served overseas.

Welcomed by Children

Fully 5000 pupils of the Bureau of Jewish Education's Schools were gathered yesterday afternoon at 4 o'clock in the grounds of Temple Mishkan Tefila School, Elm Hill av, Roxbury, to welcome the commission. The building was handsomely decorated and the children and their elders crowded it to the street. All carried American and Zionist flags.

Rabbi H. H. Rubenovitz of Temple Mishkan Tefila welcomed the guests. The other speakers were Dr Ben Zion Mossinsohn, Dr Leon S. Medalia and Louis Hurwich. Miss Edith Belin of the Grace Aguilar Religious School of the West End presented the guests, in the name of the children gathered, a floral piece bearing the Shield of David in white and blue flowers and in her address pledged the allegiance of all the children present.

Early last evening a service was held at the Adath Jeshurun Synagogue, Blue Hill av, Roxbury, which was attended by the largest crowd that has ever been in this large edifice. The service was a Mincha service, known as the twilight service. Ezra Goldberg delivered a short address and stated that the congregation had subscribed $1000 toward the Palestine Foundation Fund.

500 at Banquet

One of the most successful kosher banquets ever held in Boston was that which took place last evening in honor of the visiting commission. Fully 500 men and women crowded the two banquet halls among them being many prominent in business and professional circles, and in educational and philanthropic organizations.

The exercises were opened by L. Selig, chairman of the banquet committee, who in turn introduced Isaac Harris as the toastmaster.

The first speaker was Prof Weizmann, who received a warm reception. Prof Weizmann declared that this was the beginning of a new epoch in the history of the Jews and that the Nations of the world were watching closely what was achieved. He told of the test confronting the Jews in making a new homeland and explained how different it was from the settling in a land already in a thriving condition.

Dr Menachim M. Ussishkin also spoke of the present conditions that prevailed in Palestine and of the need of concentrated effort in behalf of the Jews of the world.

Dr Ben Zion Mossinsohn also spoke. Prof Albert Einstein told of the institutional work that was to be carried on in Palestine. He indorsed many of the things previously told by Prof Weizmann and asked that the movement receive the indorsement that was due it.

Rabbi H. R. Gold of the Adath Jeshurun Congregation made a most earnest appeal for monetary aid for the movement. He stated that with wholesale slaughter of the Jews in Eastern Europe and with the closing of the doors of this country, the only hope for the Jew was in Palestine. He drew a most graphic picture of the suffering of the Jews in Europe and told of the great opportunities that were presented by the upbuilding of Palestine. The appeal was answered by the subscription of a very large sum, which will be announced by the committee in charge today.

Another Busy Day

The visitors will be kept busy from early this morning until late tonight. Their program for the day follows:

> 10 AM—Reception by Pres Lowell at Harvard University.
> 1 PM—Luncheon tendered the visitors by Gov Cox at Hotel Touraine, followed by official greeting of State at State House.
> 4 PM—Reception to Mrs Einstein, Mrs Weizmann and ladies by Boston Zionist Women at Mishkan Tefila Temple.
> 4 PM—American Academy of Arts and Sciences reception to Prof Einstein at Cambridge.
> 5:30 PM—Public reception and demon[not continued] [Dinner given by New Cen]tury Club at Boston City Club.
> 8 PM—Public reception and demonstration at Boston Opera House.
> 8 PM—Public reception and demonstration at Mechanic's Building. Members of the commission will alternate in addressing both meetings.

The Boston Daily Globe, May 18, pp. 1, 6.

Albert N. Einstein? True, this Albert was *an* Einstein but what was this curious spelling good for?

Wednesday, May 18

Einstein Visits Harvard

State Luncheon in His Honor at Noon and Several Events for the Afternoon and Evening

Before going to Hotel Touraine this noon for the State luncheon, given in his honor by Governor Cox, Professor Albert N. Einstein and his wife went to Harvard College. They were received first by Julian L. Coolidge, professor of mathematics, and John U. Nef, graduate manager of the Harvard Union, and presented to President Lowell in the faculty room in University Hall. President Lowell greeted Dr. Einstein in French and introduced him to members of the faculty, including Theodore Lyman, professor of physics; Theodore W. Richards, professor of chemistry, and former Nobel prize winner; William Duane, professor of bio-physics and authority on the subject of X-rays; Charles H. Haskins, dean of the graduate school of arts and sciences; Roscoe Pound, dean of the law school; Percy W. Bridgman, professor of physics and Professor MacMechan of Dalhousie University, who was visiting Harvard.

Then Professor Lyman and Professor Duane escorted their guest to the different buildings, museums, classrooms and laboratories which he expressed a desire to see. There were no formalities and no speeches.

The guests at the luncheon given by Governor Cox to Professor Einstein and Professor Weizmann were as follows:

President Lemuel Murlin, President Henry Lefavour, Elihu Thompson, President John A. Cousens, Dr. Payson Smith, Adolph Ehrlich, Ferdinand Strauss, David A. Lourie, A. K. Cohen, A. C. Ratshesky, Elihu D. Stone, Jacob Wiseman, Solomon Lewenberg, Alexander Brin, Professor Milton Roseneau, Maurice E. Goldberg, Jacob Asher, Sidney Conrad, Ernest J. Goulston, Adjutant General Jesse P. Stevens, Isaac Harris, Albert Hurwitz, Carl Dreyfus, Herman A. MacDonald, Charles M. Stiller, Hon. Alvan T. Fuller, Charles M. Southworth, Samuel W. McCall, Eugene N. Foss, Frederic A. Junisky, Joseph Talamo, Charles Shulman, Bernard Finkelstein, Harry H. Williams, Horas A. Carter, William J. Foley, John C. F. Slayton, James F. Ingraham, Charles S. Smith, John A. White, Francis W. Aldrich, M. Ussishkin, Dr. B. B. Messinsohn, Mrs. Einstein, Mrs. Weizmann.

For the remainder of the day and evening the programme is as follows:

4 P.M.—Reception to Mrs. Einstein, Mrs. Weizmann and others by
Boston Zionist Women at Mishkan Tefila Temple.

4 P.M.—Reception to Professor Einstein at American Academy of Arts
and Sciences.

6.30 P.M.—Dinner by New Century Club at Boston City Club.

8 P.M.—Simultaneous public receptions and mass meetings at Boston
Opera House and Mechanics Building.

The Boston Evening Transcript, May 18, p. 5.

Einstein had been invited to visit at Harvard, in the name of President
Lowell, by Theodore Lyman, and finally by Lowell himself on May
11. Upon accepting the invitation on May 4, Einstein named T. Harris of 5–7 Beacon Street as the person responsible for his academic
program in Boston.[70]

When he visited Harvard, "Professor Theodore Lyman, famous for
his optical investigations, informed him about the work that was being done there, . . ." Philipp Frank adds. "Einstein . . . allowed several
students to give him detailed explanations of the problems on which
they were working. Furthermore he actually thought about these
problems, and some students received advice from him that was helpful in their research."[71] Lyman was the first American physicist who
proposed Einstein for the Nobel Prize.[72]

Einstein did not deliver lectures at Harvard—this was noticed by
certain members of the Zionist delegation. They suspected interference from the leadership of the Zionist Organization of America
(ZOA). Felix Frankfurter, professor of law at Harvard and one of the
leaders of the ZOA, felt it necessary to deny the accusation.[73]

The invitation of the Harvard Union is not available.

He was also invited by the Harvard Student Liberal Club for a talk
on international understanding.[74] The club let it be known that it was
looking to Einstein for leadership and that it expected a great impetus to the intercollegiate liberal movement from Einstein's visit. The
invitation does not mention Einstein's scientific achievements; he is
approached as an international personality, a potential leader.

Probably pressed for time, Einstein missed an interesting experience by not meeting with the Club.

Prof. Einstein Welcomed by Harvard Head

Visits Observatory with Party Early Today

Prof. and Mrs. Albert Einstein and Dr. and Mrs. Chaim Weizmann began the second day of their Boston visit with a morning motor ride to Cambridge, where Pres. Lowell of Harvard gave them a semi-formal reception. Prof. Einstein was much interested in the university, partly because it includes the Harvard Astronomical Observatory. His theory of "relativity" has a great deal to do with the study of the stars and measurement of space and time.

The visitors had to hurry back to Boston to take luncheon with Gov. Cox at the Touraine at 1 o'clock. Their afternoon schedule included two receptions, at 4 o'clock one given by the American Academy of Arts and Sciences at Cambridge for Prof. Einstein and Dr. Weizmann, the other given for their wives by Boston Zionist women at the Mishkan Tefila Temple grounds.

The visitors faced a strenuous evening, beginning with a dinner by the New Century Club at the Boston City Club, and ending with mass meetings in the Boston Opera House and Mechanics Building.

Record, May 18.

> My apologies for the possibly incomplete title of the newspaper. This clipping from the Albert Einstein Archive does not have enough details to identify it. Any help would be welcome.
>
> His lecture at the American Academy of Arts and Sciences was an "admirably lucid exposition of the comprehensive theory of relativity," as President Moore formulated it[75]—a platitude, but significant for being the only trace of Einstein's having been there at all.

Guests at Dinner

Members of Commission Honored by New Century Club

The New Century Club, an organization of Jewish professional men who are interested in things that help maintain high standard among their people, entertained the members of the Weizmann-Einstein commission last evening at a banquet given at the Boston City Club. The guests of honor, although they were all scheduled to speak at the Boston Opera House and

The American Academy of Arts and Sciences, Boston, 28 Newbury Street, where Einstein delivered his lecture to the Academy.

the Mechanics building demonstrations, divided their forces, so that they continued to address their Boston hosts until a late hour.

Nearly 500 members and guests were at the dinner, at which the president of the club, Philip Pinckney, introduced David A. Lourie, who presided as toastmaster. The distinguished guests, said Mr. Lourie, represent an ideal of the Jewish people which but yesterday seemed but a fantastic dream, and yet today it is coming to pass as a practical reality. The New Century Club, he said, is interested in the cultural side of life, the development of which is part of the Zionistic work in Palestine and the Hebrew tongue, the language of the Bible, he said, is to be made a living force in the world. He paid a tribute to the learning of Dr. Chaim Weizmann, whose genius had given to the allies a force which had helped materially in winning the war. For this he had asked and had received no material reward, except the granting of recognition to his own people.

Dr. Weizmann dwelt upon the homelessness of the Jewish people and the fact that their genius has gone to swell the aggregate of the achievements of any people, among whom they have dwelt.

Dr. Einstein spoke briefly upon the aspirations of the Jewish people for a national existence of their own.

The Boston Herald, May 19, p. 14.

> The promise to turn Palestine into a Jewish National Home was, of course, not a reward for Weizmann's war service but the result of the hard political work of Zionists, doing their best to convince British politicians, among them British Jews, of the feasibility of resettling Palestine.

Respond to Appeal

Members of New Century Club Give $25,000 toward New University in Palestine—Reception to Jewish Leaders

With unbounded enthusiasm more than $25,000 was raised last night at the dinner of the New Century Club, held at the City Club, for the benefit of the Hebrew University fund. Professor Albert N. Einstein and Professor Chaim Weizmann being guests of honor. Subscriptions started with a rush of $500 donations and went forward so rapidly that it was with difficulty the secretaries were able to write the names and amounts called out.

It was nearly 11 o'clock last night before Professor Einstein was able to leave the City Club for his other speaking engagements of the evening. Before he departed, before and after his brief speech, he was cheered continuously for several minutes.

Dr. Einstein's day began with a reception at Harvard, where he was escorted about the university by President Lowell and others, and included a luncheon in his honor at the Hotel Touraine, given by Governor Cox, a talk of over an hour's duration on his theory of relativity at the American Academy of Arts and Sciences, and public demonstrations on a huge scale at the Boston Opera House and the Mechanics Building late last night.

David A. Lourie, prominent in business circles in Boston, and chairman of the entertainment committee of the New Century Club, presided at the City Club dinner last night given at the City Club.

Professor Weizmann, in his plea for the cause of the great Hebrew university that is to be established, was the recipient of much applause as he made point after point.

Mendell Ussichkin of Palestine made the next appeal for the cause of the university, being frequently interrupted by applause.

Dr. Ben Mossinsohn spoke at length on the ideals of a great Hebrew university. In the past he said that the Jew had only the choice of two things—to either make money or become a scientist—and the last road was practically closed. Now, with the coming of the new university, will come a changed cultural era. He urged the need of bringing the Jewish people to Jewish soil.

Then Mr. Lourie addressed the 400 members of the club present, saying that for 21 years it had been an inflexible rule of that organization that no funds should be solicited at club gatherings. But as this was a time of crisis the rule was to be ignored for the time and contributions for the Hebrew university asked.

Before departing for his other engagements Dr. Einstein, speaking in German, addressed the club members. He said that if the Jews have a university in Palestine they will have a culture identified with the Jewish people. He added that if Jews live in Palestine they can gain a freedom of expression that would be impossible otherwise. He said that this cannot be accomplished as individuals, but must be done by a large and comprehensive group.

The Boston Post, May 19, p. 12.

Why just the New Century Club? The reason for it lies, again, in the divisions within American Jewry. The most exclusive in Boston was the Elysium Club, organized by "German" Jewish leaders, while the New Century Club was formed by Eastern European Jews. The Elysium always remained dominated by "Westerners," and the New Century became the Russian social enclave.[76] It was obvious, therefore, that Einstein, as a member of a delegation consisting only of Easterners, all from Russia, was invited by this club and not by the other.

Appeals to Jews for Funds

Weizmann and Einstein Wildly Greeted—Women Give Jewels to Help the Cause—More Than $150,000 Donated for Palestine

Stirring appeals to the Jews of New England to submit to "self-taxation" that sufficient funds might be raised to realize the Zionist ideal of a reconstructed Palestine, were made last night at three big meetings in this city, by leaders of world Jewry.

Thousands of Jews assembled at the Boston Opera House, Mechanics' building and at the New Century Club dinner at the City Club responded generously to these appeals by contributing more than $150,000 to the world movement to raise $100,000,000 for the rebuilding of Palestine.

The appearance of the speakers, Dr. Chaim Weizmann, president of the World Zionist organization, Professor Albert Einstein, noted scientist, Dr. Ben Zion Mossinsohn, principal of the Hebrew University in Jerusalem, and M. M. Ussischkin, head of the Zionist Commission in Palestine, at the three meetings brought forth thunderous applause.

Opera House Crowded

A throng which filled the Boston Opera House paid tribute to the leaders of the World Zionists. Isaac Harris, chairman of the committee, who presided over the meeting, presented the noted guests to the thousands of representatives of American Jewry present—Zionists and non-Zionists alike.

Dr. Weizmann, when he arose to speak, received an ovation. The audience arose, waved American and Zionist flags, and sang "America" and "Hatikvoh," the Zionist national hymn.

"Jewish honor and perhaps Jewish existence is at stake and the key to

Palestine is not in the pocket of Sir Herbert Samuels, British high commissioner in Palestine, nor of the Zionist movement, but in the pocket of the American Jew," he said.

"There are three lines upon which the uplifting of Palestine should proceed," Dr. Weizmann said. "The Jewish national home and the Jewish commonwealth which is to follow, should stand on the land of Palestine. There is land sufficient in Palestine to sustain a population eight to nine times greater than is there now.

"There must be men and women ready to take over the land and transform it into what it was thousands of years ago, or what can be made of it through modern agricultural methods. As soon as possible there should be a vast number of Jews in Palestine to develop the land.

"The possibilities of educational development must be considered. We ask for means to establish a Jewish university to keep alive the intellectual power which has kept us alive these thousands of years."

He urged all Jews to aid to the point of sacrifice in upbuilding Palestine, and said those in this country should send not only money, but young men and women as well, to help in the work. He urged all Jews alike to do all they could for the movement that Palestine might become the centre of Jewish life.

"Our Jews in Boston will respond," Chairman Harris told the Zionist leaders. Then began the flow of money from all sections of the Opera House. Contributions upon contributions were announced. Harris Selig, of Boston, also urged in the appeal. In the space of a few minutes it was announced that thousands of dollars had been donated.

Homeland to Jews

Dr. Ben Zion Mossinsohn in his address said: "We are living in a solemn moment in the life of the Jewish people. For 2000 years they have prayed and worked that this day might come to pass, and, through the miracle following the world chaos, here is rising therefrom the restored Palestine as the homeland again to the Jewish people."

Chairman Harris said the age-old dream of the Jew is about to be realized. We are privileged to be the living witnesses to the greatest drama in the history of the Jew. "Today there comes to us with his associates the man who more than any other individual deserves the credit for achieving the prayers of the Jewish people. We are here to celebrate their being with us, for the first time in history, the chosen leader of the Zionist organization of the world."

Dr. Ben Zion Mossinsohn when presented to the audience received a great ovation, and one again the audience sang the Zionist anthem.

Dr. Mossinsohn said the Jewish people must face squarely the tasks before them. "In brief and simple words every Jew is expected to do his utmost toward the upbuilding of Palestine not only as a refuge of the oppressed and poor, but as the home of the Jewish people. And because it is going to be the home for those who want it to be their home, we shall have to do our very best, and our very best is just good enough for Palestine.

"Your assistance will be the answer to all this flag waving. Along with the waving of flags and the singing, another din rings in my ears—the tramp of the Chaluzim and the breaking of stones on Palestinian roads in order to fulfil this great duty."

Menachem M. Ussischkin, head of the Zionist Commission in Palestine, spoke along similar lines in Yiddish.

The Boston Post, May 19, pp. 1, 12.

Jewish Women Give Heap of Jewels to Aid Palestine Fund

Commission Has Already Raised $150,000 Here—Two Great Meetings Held in Opera House and Mechanics Building—Both Gatherings Hear Noted Leaders—Are Guests of Gov. Cox at State Luncheon—Dr. Einstein Visits Harvard

Jewish women of Boston are literally "giving until it hurts" in obedience to Dr. Chaim Weizmann's principle of "self-taxation." They are stripping the rings from their fingers, bracelets and wristwatches from their arms, pendants from their ears and casting lockets, chains and necklaces, in many instances heirlooms, into the "jewel pot" for the benefit of the Palestine Foundation Fund, which is to reconstruct the Holy Land.

Last evening from the stages of the Boston Opera House and of Mechanics building, it was announced that the combined gifts of Boston Jewry in the last two days to the foundation fund, as the result of the appeal made by the Weizmann-Einstein commission, aggregated more than $150,000, and the harvest has not yet been completed.

Opera House Sold Out

The announcement was made by Isaac Harris, chairman of the meeting in the Opera House, and by Harris Selig, who voiced the appeal. The Opera House was sold out to prospective givers, at from $1 to $3 a seat, while the main hall of Mechanics building was about half filled at slightly lower prices.

Dr. Albert Einstein—the scholarly physicist of Berlin and more recently of the University of Leyden, and the enunciator of the famous doctrine of "relativity"—and Dr. Weizmann of the University of Manchester, the discoverer of the terribly destructive power of trinitrotoluol—TNT—were the principal speakers.

Dr. Einstein spoke only in German. Dr. Weizmann spoke in English and depicted the deplorable conditions in Europe and predicted the calamities that will result to the Jews, as well as the other peoples, if the half million Jewish orphans in the Ukraine are not cared for, and other remedial work undertaken, for which the reconstruction of the Holy Land provides a practical basis.

Other speakers at the various meetings held yesterday were Menachem Mendel Ussischkin, a civil engineer from Palestine and a member of the executive committee of the World Zionist Commission, Dr. Schmaryu Levin and Dr. Ben-Zion Mossenssohn.

Prof. Einstein, accompanied by Mrs. Einstein and by his secretary, Solomon Ginsberg, were taken to Harvard University in President Lowell's automobile yesterday forenoon. They were escorted by Julian L. Coolidge, professor of mathematics, and John U. Nef, graduate manager of the Harvard Union. At University Hall they were presented to President Lowell, who greeted Dr. Einstein in French and introduced him to members of the faculty.

At 1 P.M. Gov. Cox was host to the visitors at a state luncheon at the Hotel Touraine. Former Governors Foss and McCall and Lt.-Gov. Fuller were among those presented to the Zionists. Heads of various universities and colleges in and near Boston and other prominent educators were guests.

Long Lecture on Theory

In the afternoon Dr. Einstein was the guest at a reception given by the American Academy of Arts and Sciences, at 28 Newbury street. Prof. George P. Moore presided. His introduction of the guest of the day was begun in

German and concluded in English, and paid tribute to Dr. Einstein's remarkable work in the realm of physics. Dr. Einstein spoke for an hour and 45 minutes, enlarging upon his famous theory. His talk was in scientific German, of which he is a master, and he illustrated it with diagrams upon a blackboard. The audience, which comprised university professors and scientific men, frequently applauded him. At the conclusion Prof. Moore thanked the noted physicist and a reception followed.

Temple Mishkan Tefila in Moreland street, Roxbury, was crowded yesterday afternoon at the reception to Mrs. Weizmann and to Mrs. Einstein. Mrs. Isaac Harris was chairman of the gathering and Mrs. H. H. Rubenovitz, wife of the Rabbi of the Temple, opened the speaking. Then Mrs. Weizmann made an appeal for the rehabilitation of the home of the Hebrew race, particularly emphasizing the world situation, in which the Jews are being slaughtered in eastern Europe and have no country of asylum to which they can flee.

The response was electrifying. Young girl ushers worked their way with difficulty through the crowded aisles, carrying long pasteboard boxes, such as are used to put up long-stemmed roses and carnations. Bills of various denominations were rained into these receptacles.

Pour Jewels Into Boxes

A prominent Boston Jewess cried out ecstatically that she also had eight sons in the war and that she wanted to make some donation in proportion to their sacrifices. She held up her watch, a valuable imported timepiece, and slipped the rings from her hands. Others followed her example and soon baskets and boxes filled with diamonds and other precious ornaments in costly settings were being carried to the front of the church. Mrs. Einstein followed Mrs. Weizmann with an address thanking the women for their response. The offering included $1000 in cash, pledges of $3000 more and sacrificial offerings of jewelry to the value of $5000.

The Boston Herald, May 19, pp. 1, 14.

The Mishkan Tefila, the city's only Conservative congregation, and Adath Jeshurun were the founding congregations of suburban Boston Jewry, the leading congregations of 44,000 Jews in Dorchester and Roxbury, the vast majority of the 47,000 living in greater Boston.

According to Mrs. Weizmann's recollections, it was she and Einstein who participated in the reception of Hadassah women. Being tired of meetings and speeches, Einstein suggested that they take a taxi and escape for some fresh air. Walking in the sun, time and again something dropped out of Einstein's hand that he picked up again and again. Curious, she asked about it. Einstein confessed that it was a tin of tooth powder. His wife Elsa did not feel well when they left New York, he explained, and he came without any luggage save a toothbrush and some tooth powder. "I never thought a great genius could be so childish!" added Vera Weizmann.[77]

I was unable to find a proper place for this story, but this seems as good a place as any. The only time Vera Weizmann met Hadassah women in Boston was her visit to the Mishkan Tefila Synagogue at 4 o'clock in the afternoon of May 18—with Mrs. Einstein, because Mr. Einstein was delivering a lecture for the American Academy of Arts and Sciences at the same time.

Jews Donate Jewelry

About $150,000 Raised in Boston during Einstein's Visit, for the Palestinian Fund—Meetings Last Night in the Boston Opera House and Mechanics Building

Boston Jews were reported last evening to have contributed about $150,000 to the fund which the Weizmann-Einstein commission has come to the United States to raise for the reconstruction and rehabilitation of Palestine. To this reconstruction belongs the establishment of the Hebrew University on the Mount of Olives.

The announcement of what has been raised was made last evening at the Boston Opera House. During the evening there was one meeting in the Opera House and another in Mechanics Building. Each was addressed by the visiting commissioners from Jerusalem. At the Opera House Isaac Harris presided and some of the speeches were in English. In Mechanics Building Rabbi Henry Raphael Gold presided, and all the speeches were in Yiddish and German. The speakers included Dr. Weizmann, Professor Einstein, M. M. Ussischkin, who is a member of the executive committee of the World Zionist Commission; Dr. Ben-Zion Mossinsohn, principal of the proposed Hebrew University and Albert Hurwitz, assistant attorney general of Massachusetts. They did not arrive at Mechanics Building until about nine o'clock, but there were about 3000 men and women waiting for them, and the speaking continued until after eleven o'clock.

Contributions took the forms of cash donations, pledges and jewelry. Boxes full of bracelets, rings, watches and pendants were carried to the platform. The audiences were told that $100,000,000 are wanted, and that America is relied upon to subscribe about three-fourths of it, to be paid at the rate of $20,000,000 a year for five years. During that period the Jews must demonstrate that they really want Palestine, or it will be colonized by England.

The Boston Evening Transcript, May 19, p. 8.

Give Their Jewels

Zionist Women of Boston Respond to Plea of Mrs. Einstein and Mrs. Weizmann by Contributing Rings, Necklaces and Earrings to Fund for Palestine

Zionist women of Boston in an unrestrained outburst of applause and acclamation at the reception yesterday afternoon to Mrs. Albert Einstein and Mrs. Chaim Weizmann, wives of the leaders of world Jewry, in the Mishkan Tefila Temple on Moreland street, Roxbury, turned what was expected to be a quiet and very formal reception into a thunderous manifestation of cheering that would have done justice to a political convention made up of men.

In a scene without precedent, at least in Boston Jewish circles, women crowding every pew and the gallery of the temple, sitting on the steps in the rear of the hall and filling the places reserved on religious occasions for the choir, ripped earrings from their holds, removed necklaces of pearls and rare stones, stripped their fingers of whatever rings they held, and in a two-hand scoop swept, in some instances, mesh bags and purses, with all their contents, into a huge box carried around by an usher to make the glittering collection.

It was impossible yesterday to make any approximate computation even of the results of this reception, to be turned over to the $100,000,000 Palestine foundation fund, but the jewels and the silver and gold trinkets and things were of sufficient quantity to fill a medium-sized traveling grip. When jewels worn by the women in the hall had been poured into the box carried about by the usher, pledges were shouted from every part of the auditorium promising sets of household silver and articles of great value easily transferable into cash. The pleas by the two visiting women were brief and simple, remarkable for their earnestness and their subdued tone.

The Boston Post, May 19, p. 12.

No English Spoken

Every Address in Mechanics' Building Either in Yiddish or in German—Much Enthusiasm Aroused by Speakers

The Mechanics' building mass meeting was opened with an address by Albert Hurwitz, assistant attorney-general for Massachusetts, who introduced Rabbi Henry Raphael Gold of the Adath Jeshurun Synagogue. Reviewing the recent history of Zionism he applauded the efforts of Dr. Weizmann in securing the Balfour declaration. Since that declaration, he said, England has proved contrary to charges that she is not "playing politics" with the Jewish people. Palestine will belong to the Jew provided the Jew responds.

Rabbi Gold introduced Menachem Mendel Ussischkin.

"It is impossible," Reb Ussischkin began in his Yiddish address, "to separate in Palestine the body from the spirit. Such feats are only permissible in America. If there are those who suggest a diminution of the budget for Jewish cultural education in Palestine they are ill with an illness so severe as to demand the most skillful of medical treatment, for," he emphasized, "they know not that they are ill.

"This is not a charity that we ask. We plead for no alms. You gave a lot of money to Esau, the war God, during the great conflict. We ask you now to speak from your heart and give to Jacob. Three-fourth of the $100,000,000 Palestine foundation fund must come from America."

Professor Chaim Weizmann, under escort of a squad of Jewish Legionnaires, headed by Sergeant-Major Jack Williams, commander of the Jewish Legion of Massachusetts, marched, cane and hat in hand, down the aisle to the platform. The scene was repeated later when Dr. Ben-Zion Mossinsohn made his entry, followed finally by the arrival of Dr. Albert Einstein.

Every address at the Mechanics' building, unlike the addresses at the Boston Opera House, was given either in Yiddish or in German, the latter language being employed by Einstein in a speech that occupied less than three minutes.

The Boston Post, May 19, p. 12.

15

Push Hebrew Medical Work

✦

Push Hebrew Medical Work

American Committee to Raise $1,000,000 for Jerusalem University

In order to make possible the establishment and the equipment of a medical department at the Hebrew University of Jerusalem, in the interests of which Professor Albert Einstein is now campaigning here, a committee of one hundred has been formed, in which many prominent Jewish physicians of the city are participating and whose purpose will be to raise $1,000,000 for the university.

Dr. Nathan Ratnoff, who initiated the movement, is president; Dr. A. M. Hilkowitch, who only a few months ago left the ranks of the British expeditionary forces in Palestine, is vice president; Dr. B. Roman and Dr. M. R. Robinson are secretaries and Dr. A. G. Rongy is treasurer.

Dr. Leo Burger is chairman of the scientific committee and Colonel J. Kopetzki is national organizer.

More than $100,000 already has been subscribed by New York physicians alone, officials of the committee announce, and the movement is bringing highly encouraging responses from all parts of the country.

Professor Albert Einstein, who came to this country in the interests of the Hebrew University, is reported to be elated with the splendid work of the American Jewish physicians' committee.

On May 21 a dinner will be tendered to him and to Professor Chaim Weitzman by the committee at the Waldorf-Astoria and not less than 1,000 physicians are expected to attend. Professor Einstein will not speak of the theory of relativity that evening, but will make public certain plans regarding the Jerusalem University which will cause no little surprise to the scientific world.

The New York Call, May 16, p. 7.

Thursday, May 19

For this day, Einstein envisaged a small gathering at 8:30 p.m. at the Hotel Commodore. The object of the gathering was to discuss with "a number of gentlemen of standing in the community" their support of the proposed Hebrew University in Jerusalem.[78] Among those invited were around fifty non-Zionists. The participants were expected to set up a provisional committee as an official partner of the Zionist Organization in university matters.[79] Of the thirty-seven responses available in the Albert Einstein Archive, only six express consent and a promise to participate.

Saturday, May 21

Jewish Doctors Raise $250,000 for College

800 at Waldorf-Astoria Hear Prof. Einstein Make Plea for University in Jerusalem

Professor Albert Einstein, Dr. Chaim Weizmann, President of the World Zionist Organization and other members of the Zionist mission to this country were the guests last night at dinner in the Waldorf-Astoria of American Jewish physicians who are trying to raise $1,000,000 to build and equip medical school of the Hebrew University at Jerusalem.

There were 800 physicians in the grand ball-room of the hotel, and it was announced they had subscribed $250,000. Dr. Nathan Ratnoff, Chairman of the Physicians Committee, presided, and said that no amount of persecution or oppression has ever checked the pursuit by his people of knowledge and culture.

"The Jewish students have always been handicapped in their pursuit of knowledge," said Prof. Einstein. "In many countries the universities and higher schools were closed to them except to a small percentage. In some of

"Caught in the Act!" At a Passover Night Seder, Weizmann sits at the head of the table. Keren Hayesod is supposed to be his wife. Einstein's wife is the Hebrew University (at the end of the table). The "22 outside of the camp"—the leaders of the Zionist Organization of America?—are hiding behind the curtain and are trying to steal the *Afikomen*, the piece of *Mazzah* usually hidden by the head of the Seder. The children have to find it, the head then gives them a present, and the Seder can continue. The *Afikomen* here is the *De'a* (the "opinion") on the Land of Israel. The three further participants sitting at the table are (from left to right) Mossinson, Levin, and Ussishkin.

The Big Stick, April 22.

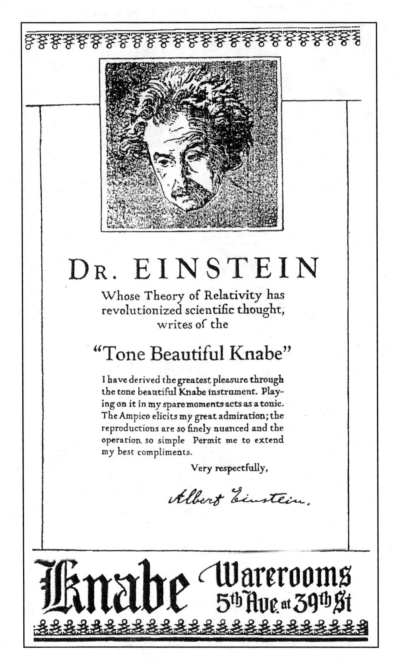

The Ampico was a nickelodeon piano. The question is whether Einstein dropped money in it—or rather, the company in Einstein's pocket.
The New York Times, April 21, p. 6.

them, while no official restrictions existed, there was a tacit understanding that no Jew should be admitted to the higher grades of professorship.

"The existence of the Jewish University in Jerusalem will give an opportunity to the Jewish mind to express himself and to receive due credit. The Medical College will undoubtedly be the most important department of the University, as we Jews have always excelled in this particular branch of science."

Other speakers were Dr. A. M. Hilkowich, Dr. Samuel J. Kopetzky, Prof. Weizmann and Dr. Schmaryah Levin.

The New York Times, May 22, p. 21.

The Little Word Adonay

Prof. Einstein and the Zionist Commission now in the country were banqueted by the American Jewish Physicians' Committee last Saturday evening. This committee is to undertake the building of the Medical Department of the proposed University in Palestine. (Some call it Hebrew, some Jewish, some Palestinian.) Behind the speakers' table was hung a painting of the University building, with the motto *Ki Mal'ah Ha-aretz Day'ah,* "For the earth shall be full of knowledge." The last word of the quotation from the Prophet is omitted. The word is *Adonay.* It is a little word, but it holds the essence of the Prophet's vision. The Prophet probably understood the importance of *knowledge* in the sciences of his day, as we do in our day. But he also understood that science was *not* the salvation of his generation, as we know that science has not been and cannot be the salvation of our day. *Day'ah Adonay,* "knowledge *of the Lord,*" was the all-important factor in the civilizing influence of the Prophet's time, as it is in our own. If, therefore, the proposed University, to be fostered by Jews, deliberately eliminates the little word *Adonay* from its motto, then it might just as well not be built. When Jews discard *Adonay,* they no longer have a *raison d'etre.*

The American Hebrew, May 27, p. 29.

16

Coming to Hartford

✦

Thursday, May 12

Plan Welcome to Zionist Chiefs

Weizmann, Ussishiken, Levin and Mossensohn Coming Here May 22

Dr. Chaim Weizmann, president of the world Zionist organization, to whom the board of aldermen has voted the freedom of the city when he comes here Sunday, May 22, will be accompanied by Menachem Mendel Ussishiken, Dr. Ben Zion Mossensohn and Dr. Schmarya Levin.

Ussishiken is a scientific engineer who is at the head of the Palestine Zionist Commission. Dr. Mossensohn is the head of the gymnasium at Jaffe. Dr. Levin is one of the deepest thinkers in present Jewish life.

The rabbis of the city will proclaim a holiday and the Jewish places of business will be closed. Thousands of Jewish children will sing "Hatikvoh" for the distinguished visitors who will arrive from New York by auto, and there will be a parade, followed by a reception to the guests at one of the leading hotels, and then a mass meeting at the Capitol theater.

The following organizations and their committees have signified their co-operation in this event:

Mizrachi, Rabbi I. S. Hurewitz, S. Levy; Poale Zion, M. Kurzmack; Hadassah, Mrs. B. Rappaport; Sharah Torah, J. Tulin; Zionist District, Joseph Hoffenberg; Workingmen's Circle, N. Promisle; Emanuel Synagogue, Walter Beatman; Agudas Achim, Harry Roth; Beth Jacob, L. Freedman; Y.M.H.A., Isaac Rudin; Y.W.H.A., Marion Brook; Sons and Daughters of Herzl, M. Paskar; Order Sons of Zion, H. B. Sisenberg; Bnai Zion Achuza, S. Glasser; Anche Koretz Synagogue, A. Kelmanson; Independent Workingmen's Circle, No. 67, M. Telechansky; Rebecca Lodge, Mrs. J. Berenson; Deborah Lodge, Mrs. Bertha Rothschild; Council of Jewish Women, Mrs. Milton Simon; Sisterhood of Emanuel Synagogue, Mrs. L. M. Schatz; Emanuel League, Miss Ethel Greenberg; Dreyfus I.O.B.S., S. Katzman; Bnai Brith, M. M. Epstein; Aaron Lodge I.O.B.S., Max Goldenthal; First Connecticut I.O.B.S., O. Dudy; Progressive Lodge, A. Kalechman.

The chairman of local organizations is Director Abraham Goldstein, who has organized the joint co-operative forces of these organizations. The entire event is under the auspices of the State Regional Zionist Union of Connecticut. The following comprise the executive of this union in charge of the event.

[Here the article suddenly ends, but because the last two paragraphs were also published in the *Hartford Daily Courant* of the same day (apparently both newspapers followed the same handout), the missing list of committee members appears in the next article.]

The Hartford Daily Times, May 12, p. 7.

> Here you have met two further fraternities: the Independent Order of B'nai B'rith (Sons of the Covenant), founded by German Jewish immigrants ("Westerners"), and the Independent Order of B'nai Sion (Sons of Sion), here represented as "I.O.B.S."

Dr. Weizmann First Jew to Receive Freedom of City

President of World Zionist Movement to Visit Hartford, May 22

The most noteworthy event in the history of Hartford Jewry is the approaching visit of the four Palestinian leaders on Sunday, May 22. For many weeks the headquarters of the Connecticut Zionist Regional Union at No. 1325 Main street has been a haven of activity and preparation for the welcome and reception to be extended to Professor Chaim Weizmann and his colleagues. Monday evening the council extended the freedom of the city to Professor Weizmann and his associates, passing a resolution which will be presented to the guests upon their arrival, an unusual precedent in itself.

President Joseph S. Holstein of the leading orthodox synagogue, has offered the use of his machine to go directly to New York and bring the guests from the Hotel Commodore to Hartford. He will be met by over 100 automobiles from various parts of the state at the Berlin turnpike. This machine procession will go on the state highway to Hartford, and parading through the principal streets, will be met by thousands of little Jewish children, appropriately clad, who will act as escorts, singing Hatikvoh, along the line of march. The Jewish homes of Hartford will be gaily decorated with American, Jewish and English flags and bunting. The rabbis of the city will proclaim a general holiday and every Jewish place of business will be closed.

A reception is to follow the parade at one of the leading hotels, after which the Jewish representatives will sit down at a dinner especially arranged by a committee. Following this dinner a mass meeting has been arranged at the Capitol Theater. The speakers will be Professor Weizmann, Menachem Mendel Ussishiken, Dr. Ben Zion Mossensohn and Schmarya Levin.

Professor Weizmann, discoverer of TNT and one of the world's geniuses is president of the World Zionist Movement. Dr. Weizmann refused compensation for his remarkable discovery of TNT from the British government, and only asked that the Jews be given a chance to once more settle in their ancient home-land, Palestine. Ussishiken is a scientific engineer who is at the head of the Palestine Zionist commission. Dr. Schmarya Levin is one of the deepest thinkers in present Jewish life and is renowned for his ability as a platform speaker. Dr. Ben Zion Mossensohn is the head of the gymnasium at Jaffre.

The following organizations and their committees have signified their cooperation in this event.—

Mizrachi Rabbi L. S. Hurewitz, S. Levy; Poale Zion. M. Kurzmack; Hadassah, Mrs. B. Rappaport; Sharah Torah, J. Tulin; Zionist District, Joseph Hoffenberg; Workingmen's Circle, N. Promisle; Emanuel Synagogue, Walter Beatman; Agudas Achim, Harry Roth; Beth Jacob, L. Freedman; Y.M.H.A., Isaac Rudin; Y.W.H.A., Marion Brook; Sons and Daughters of Herzl, M. Paskar; Order Sons of Zion, H. B. Sisenberg; Bnai Zion Achuza, S. Glasser; Anche Koretz Synagogue, A. Kelmanson; Independent Workingmen's Circle, No. 67, M. Telechansky; Rebecca Lodge, Mrs. J. Berenson; Deborah Lodge, Mrs. Bertha Rothschild; Council of Jewish Women, Mrs. Milton Simon; Sisterhood of Emanuel Synagogue, Mrs. L. M. Schatz; Emanuel League, Miss Ethel Greenberg; Dreyfus I.O.B.S., S. Katzman; Bnai Brith, M. M. Epstein; Aaron Lodge I.O.B.S., Max Goldenthal; First Connecticut I.O.B.S., P. Dudy; Progressive Lodge, A. Kalechman.

Notable in the activities of the above organizations is the work of the

Y.W.H.A. and the Hadassah, who are manifesting able and enthusiastic support and will furnish all the necessary ushers for the Capitol Theater meeting.

The chairman of local organizations is Director Abraham Goldstein, who has organized the joint co-operative forces of these organizations. The entire event is under the auspices of the State Regional Zionist Union of Connecticut. The following comprise the executive committee of this union in charge of the event.

Morris Cohen and S. Kaplan, New Britain; Oscar Frank, Meriden; B. Press, Middletown; Avner Schwartz, Norwich; A. Rosenberg, Ellington; A. Israel, Willimantic; Samuel C. Kone, Reuben Taylor and Leo Glassman, Hartford; B. Epstein, Winsted, Mrs. A. Cohen, Bristol.

Among other committees are the following:—

Dinner, Joseph S. Silver, chairman, Ellik Nirenstein, S. Hershman, J. R. N. Cohen, Harry Levin, I. E. Goldberg, ex officio: Abraham Goldstein.

Executive of the reception committee, I. Goldberg, chairman; Joseph S. Silver, Samuel Hershman, L. P. Toft, William Rulnick, Walter Beatman, Harry Levin, J. Mintz, E. Nirenstein. Louis Feinberg, Joseph S. Holstein, George Levine, Rabbi Cemach Hoffenberg, Rabbi Isaac Hurwitz, Rabbi Abraham Nowak, Reuben Taylor, H. P. Koppleman, George S. Schwartz, Senator Louis Rosenfeld, S. D. Tulin, Samuel C. Kone.

Publicity, Samuel C. Kone, Leo Glassman, A. Nevelstein, A. Marder.

The reception committee will consist of about 300 of the most prominent Jews of Connecticut and will be formally mentioned next week.

Tickets for the mass meeting are on sale at the Zionist Headquarters, No 1,325 Main street, and may be obtained both day and evening.

The Hartford Daily Courant, May 12, p. 5.

Monday, May 16

Zionists Plan for Weizmann Welcome

Leading Jewish Societies to Take Part in Event—Bells Will Play National Anthem—Churches to Aid in Honoring Visitors Next Sunday

One of the features of Connecticut's welcome to the four Zionist leaders next Sunday, at Hartford, will be the chiming of the national anthem "Hatikvoh" from the tower of the principal Protestant church on Main street and the belfry of the Catholic cathedral on Farmington avenue. The Talmud Torah children will be clad in blue and white, emblematic of the colors of the Jewish flag now floating over Jerusalem in honor of the event.

The message of Chaim Weizmann, Ussishkin, Levin and Mossessohn to the Jews of Hartford will deal with vital problems connected with Palestine. There is land to be bought and fertilized. There are roads and railways, harbors, bridges, viaducts, cities and towns to be built. There is soil to be irrigated. There is latent water power to be harnessed. There are houses to be built, streets to be paved, sanitary systems to be introduced and avenues to be laid out.

I. E. Goldberg, chairman of the reception committee, reports that the committee is being increased daily by the returns that reach his office by those who are to be on this committee which follows:—

I. E. Goldberg, chairman; Joseph Silver, Rabbi Nowak, Rabbi Hoffenberg. Rabbi Hurewitz, Rabbi Anspacher, S. D. Tulin, Hyman Kaplan, A. Goldstein, L. P. Toft, S. Hershman, William Rulnick, Harry Levin, James Mintz, Walter Beatman, Barney Rapaport. E. Nirenstein, Joseph Holstein, Wolf Silver, Jacob Silver, B. Wachtel, L. S. Knoek, Saul Berman, H. P. Koppleman, Solomon Elsner, Albert M. Simons, Senator Rosenfeld, Louis Kemler, M. L. Silverman, Ralph Kolodney, William Vogel, D. A. Rosen, F. Rabinowitz, N. Roth, Samuel E. Herrup, A. Katten, S. H. Leavitt, William Raphael, S. C. Kone, A. S. Borden, Samuel S. Mattes, S. S. Tulin, Reuben Taylor, Jos Hoffenberg, I. Paul, J. R. N. Cohen, A. Hoffenberg, M. S. Sheketoff, I. Silver, Abraham Hoffman, David Kaplan, M. Juster, H. Krassnow, Louis Y. Gaberman, David Traub, H. I. Epstein, J. Shenker, Alex Tulin, H. Perstein, Louis Kaplan, S. Hoffenberg, L. L. Alexander, George B. Schwartz.

Every Jewish organization will have a representative on the reception committee. Names are to be sent in to the headquarters not later than Tuesday. A meeting of the ushers from the Hadassah, Y.W.H.A., Sisterhood of Immanuel Synagogue, Rebecca Lodge, and other women's organizations will take place tomorrow night at the Zionist Headquarters No. 1325 Main street, where all final instructions for the mass meeting will be concluded. A. Goldstein announces that a final meeting of all the committees arranging this affair, and a large number of members of the various organizations who are also participating, will take place Wednesday night, at the Emmanuel Synagogue vestry room. Reports will be given as to all the details.

The Hartford Daily Courant, May 16, p. 3.

Tuesday, May 17

Einstein Coming to Hartford Sunday with Other Zionist Leaders

Professor Einstein, who came to this country recently for the cause of the Hebrew university in Jerusalem, will visit Hartford, Sunday, in company with Dr. Chaim Weizmann, president of the World Zionist organization; Dr. Schmarya Levin, a former member of the Russian duma, and M. Ussuskin, a Russian engineer who is the head of the Palestine commission.

Announcement of Professor Einstein's coming here was made to-day by Abraham Goldstein, director of Zionist activities, who has just returned from New York where he was in conference with the Zionist chiefs.

The party will arrive by auto about 10 o'clock, daylight saving time, and will be met near Goodwin park by parade in which there will be two or three hundred autos. The guests will be given a reception at the Hotel Garde and in the afternoon they will address a mass meeting in the Capitol theater.

Jewish homes are being decorated with flags in preparation for the event.

Delegates of all the local organizations co-operating in the event will meet Wednesday night at the Emanuel synagogue for final arrangements.

The Hartford Daily Times, May 17, p. 1.

Friday, May 20

Raphael Marshal of Zionist Parade

Capt. Tulin to Be Chairman of Mass Meeting in Honor of Zionist Chiefs at Capitol Theater Sunday

Arrangements for the Weizmann-Einstein reception, Sunday, have been practically completed. A committee of prominent Jews will bring the guests at 9:30 to Goodwin park, and will be followed into the city by more than 200 automobiles. Alderman Raphael, who is marshal of the parade, with the assistance of the Y.M.H.A. has arranged to have a brass band and a platoon of police march in from the South Green to the Zionist headquarters opposite the Keney Tower. An automobile has been provided for the rabbis of the city who will lead the automobile parade. The children of the

Hebrew schools and of other Jewish organizations will greet the guests at the Keney Tower and make a presentation. Miss Rae Podnetsky and Samuel Alkow, principal of the Talmud Torah, are in charge of organizing the children's reception to the guests. The guests will then proceed to the Garde hotel, where a reception will be held from 11 to 12. Stores along the line of march will be decorated. Rabbi A. Nowak, H. P. Koppleman and Alderman Raphael were the committee which arranged for the decoration of the stores.

Joseph B. Silver, chairman of the dinner, reports that all arrangements for the dinner are complete, and 175 from Hartford and surrounding towns will dine with the guests. Judge Solomon Elsner will be toastmaster.

A. Goldstein announces that the committee has chosen the well known Zionist leader, Captain Abraham Tulin of New York, to be chairman of the mass meeting, to be held in the Capitol theater. Mr. Tulin will arrive in Hartford to-night.

I. E. Goldberg, chairman of the reception committee, announced at a meeting Thursday night that the program for the reception had been completed and his plans for the Keren Hayesod campaign met with great enthusiasm.

Samuel Hoffenberg, chairman of the theater committee, and Mrs. J. R. N. Cohen, who are in charge of the floor committee, announce that seventy-five women and girls have volunteered to assist in the theater.

Special meetings are called in all the synagogues in connection with this event for Saturday night.

It is the first time that leaders of world Jewry have honored this community. This is the fourth city in the United States they are visiting, the others being New York, Boston and Chicago. It has been said that this is being done to show the appreciation of the Zionist record the Jews of Connecticut have made, and the interest they have taken in the upbuilding of Palestine.

The Hartford Daily Times, May 20, p. 2.

The same day Hartford Zionist leader A. Goldstein sent an SOS message to Weizmann: If Einstein would not come—apparently he was reluctant to visit there—the whole campaign in Connecticut would collapse.[80]

Saturday, May 21

Zionist Reception Plans Completed

Hartford Will Honor Einstein and Weitzman Tomorrow—Goodwin Park to Be Meeting Place—Platoon of Police Will Escort Scientists to Zionist Headquarters

An elaborate reception, the plans of which have been practically completed, has been arranged for tomorrow by local Jewish organizations and prominent Jewish citizens to herald the visit of Professor Chaim Weitzmann and Professor Einstein, protagonist of the theory of relativity. A committee will bring the well known guests to Goodwin Park at 8:30 a.m., and they will then be conveyed into the city by automobile, followed by a procession of more than 200 machines. A brass band and a platoon of police will join the parade at the South Green, and will lead the delegation to the Zionist headquarters opposite Keney Tower.—

One of the special features attending Prof. Weitzmann's tour through the country has been a reception given him by the Jewish children, and this will also be a feature of his Hartford visit. A committee is mustering the children of Hartford Hebrew School, Temple Emanuel and the Young Judaeans and various Jewish clubs. All the children will assemble in front of Keney Tower, at [8?] a.m. tomorrow. When the distinguished visitors will reach Keney Tower two children, a boy and girl, will present a bouquet of flowers to them.

The guests will then go to the Garde Hotel, where a reception will be held from 10 to 11. Stores will be decorated along the line of march, in accordance with the arrangements made by Rabbi A. Nowak, H. P. Kopplemann and Alderman Raphael of the committee. About 175 guests will be present from Hartford and surrounding towns at the dinner. Judge Solomon Elsner will act as toastmaster.

A message addressed to A. Goldstein was received last night at the Zionist headquarters from Mayor Brainard informing the Weitzmann-Einstein reception committee that an official of the city has been delegated to make a formal presentation of the freedom of Hartford.

Among the men who will appear on the stage of the Capitol Theater tomorrow will be Captain Abe Tulin, a Hartford man who is practicing law in New York. He was a United States Army captain during the World War and later, during the peace negotiations in Paris, was a member of an inter-allied diplomatic commission to the Near East and Greece.

I. E. Goldberg, chairman of the reception committee, announced at a meeting Thursday night that all plans had been completed for the reception. Seventy-five women and girls have volunteered to assist in ushering at the theater; according to Samuel Hoffenberg, chairman of the committee, and Mrs. J. R. N. Cohen, in charge of the floor committee.

The Hartford Daily Courant, May 21, p. 13.

Raphael to Welcome Zionist Chiefs for City

Alderman William Raphael has been delegated to make formal presentation of the freedom of the city to Dr. Chaim Weizmann, president of the World Zionist organization, Professor Albert Einstein, the noted scientist, and other Zionist chiefs who are coming to Hartford Sunday. They will arrive by auto from New Haven and will be greeted at Goodwin park, and from there escorted to the city by an automobile parade.

A reception and banquet will be held at the Hotel Garde and in the afternoon the guests will address a mass meeting at the Capitol theater.

Main street will be decorated with flags in honor of the distinguished visitors.

The Hartford Daily Times, May 21, p. 5.

Weizmann Welcome Plans Completed

Leading Jewish Citizens to Greet Famous Scientists—Zionist Flags are in Great Demand—Stores Along Line of March Decorated for Today

At a meeting held in the Zionist headquarters at No. 1,325 Main street last night, final polish was put on the arrangements for the grand reception to be given today to Professors Weizmann and Einstein and the visiting World Zionist delegation, to which the freedom of the city is to be given, A. Goldstein presiding over the meeting, at which all of the Jewish bodies of the city were represented. Improvements were outlined and the latest reports involve minor changes in the time card. Alderman William Raphael is to act as marshal of the long procession which will honor the distinguished visitors. From the report received it is expected that a large sum in cash will be presented at the mass meeting in Capitol Theater, as a contribution to the Zionist fund.

Leaving New York City at 8:45, the delegates' party will arrive at Berlin at 9:45. Its personnel will be Professor Chaim Weizmann and Mrs. Weizmann, Professor Albert Einstein and Mrs. Einstein, M. M. Ussishkin of Russia, the engineer who is head of the Palestine commission, Dr. S. Ch. Levin of Russia, who was a member of the Duma in 1905, and Secretary Ginzburg of the delegation. At about 9 o'clock this morning the reception committee will start for Berlin to meet the party at the station. J. E. Goldberg and fourteen others will be on this body. They will make the trip by automobiles and will convey the visitors from the station to Goodwin Park, where they will be due about 10:15. At the park, they will find the long column of automobiles, which is to compose the main part of the procession later to traverse the heart of the city in honor of the party.

At Barnard Park a platoon of police and a band will await the leading automobiles. The procession will start about 10:30 with Alderman Raphael as marshal. In the first machine will ride the four rabbis of the city, Rabbis S. Hoffenberg, L. Hurwitz, A. Nowak and A. Anspacher. In two machines the delegates will follow. Between 200 and 250 automobiles are to drive and then a body of Jewish citizens may march. This part of the procession will be spontaneous.

Passing the Zionist headquarters, the right of the column will halt nearly opposite Keney Tower and there a large body of Jewish children will be ready to greet the delegates. They will be from the Hartford Hebrew School, the Temple Emanuel, the Young Judeans and other bodies. A boy and a girl will present flowers to the delegates. Resuming the parade at 11 o'clock the escort will traverse High street to the Hotel Garde, where it will disband. There will be a brief interval and then, about 11:30, dinner will be served.

Judge Solomon Elsner will act as toastmaster. Professors Weizmann and Einstein, Mr. Ussishkin and Dr. Levin will speak. The speaking is to conclude at 1:45, as the exercises at the theater are to begin at 2 o'clock sharp.

Members of the delegation will speak at the mass meeting in the theater, over which Alderman Raphael will preside. The alderman will represent Mayor Brainard, who is out of the city. He will present the freedom of the city to the delegates. It is expected that a very large sum in cash will be presented to the delegation for the Palestine Foundation Fund. It is believed that the amount will astonish many. Captain Abe Tulin, who served in the army in the war will be among those on the stage. The meeting will conclude at 5 o'clock, when the delegates will be taken to a luncheon. About 6 o'clock the visitors will start for New Haven, where they are to speak this evening.

Decorating for Guests

Decorations were displayed yesterday in honor of Prof. Chaim Weizman and Einstein. The great department stores on Main street, Wise, Smith & Co., Brown, Thomson & Co., Sage, Allen & Co., and G. Fox Co., early displayed the American flag and the six-pointed star of Zion. The Zion banner has a white field crossed by two blue bands between which the star is placed.

The Hartford-Aetna National Bank decorated in the morning. The Outlet Millinery Store and other were early with flags. When evening came the section of Main street, between State and Talcott, was ablaze with numerous flags. For two days there has been a demand for the beautiful Jewish banner and it culminated with a rush in several of the stores yesterday. There was especially a call for small sizes.

The Hartford Daily Courant, May 22, p. 18.

Sunday, May 22

Zionist Chiefs Acclaimed in Hartford; Crowds Struggle to See Weizmann and Einstein

Dr. Chaim Weizmann, the man who brought about the Balfour declaration promising England's support to the Jews in their aspirations for a homeland in Palestine, came to Hartford Sunday and was wildly acclaimed. He was accompanied by Prof. Albert Einstein, the noted scientist, and Dr. Schmarya Levine, poet, philosopher and one-time member of the Russian Duma.

Resolutions of the board of aldermen, extending the freedom of the city to the distinguished visitors, were presented to them by Alderman William Raphael at a mass meeting in the Capitol theater. Nearly $100,000 was pledged among those at the meeting for the Keren Hayesod (Palestine foundation fund) and a $25,000 check was presented in advance to Dr. Weizmann.

The meeting in the theater was preceded by a reception and luncheon at the Hotel Garde.

A parade in which there were probably more than 100 automobiles was one of the interesting events of the day. It started from Goodwin park, where the guests were met upon their arrival by auto from New Haven. Motorcycle policemen made way for the procession which was headed by Colt's band and five war veterans who carried American, British and Zionist flags. The flag bearers were Max Kowalsky, Jacob Robins and Sam Chudowsky, who served in the Jewish legion under General Allenby in Palestine; B. H. Reicher, who served in the American army and Morris Tulin, a sailor.

The automobiles, many of which were from other cities in the state, were decorated with flags. David Baron was marshal. Abraham Goldstein, director of Zionist activities in Connecticut, rode in the marshal's car. The second auto in line carried the honored guests, also Alderman Raphael and S. A. Tulin of this city, who went to school in Russia with Dr. Weizmann when they were boys.

Children Present Flowers

The parade route was from Goodwin park to Main street by way of Maple avenue, from North Main street to the Garde on Asylum street by way of High street. Business houses along the way were decorated with American and Zionist banners. The honored guests were cheered all the way from automobiles and from crowds on the sidewalks. North Main street was jammed by crowds that struggled to get close to shake hands with the distinguished visitors when their car stopped for a few moments near Keney Tower.

Several hundred children of the Talmud Torah, Emmanuel Synagogue and other Young Judea societies greeted the visitors at that point with "Hatikvoh," the Zionist anthem. Isadore Levin, 12 years old, representing the Young Judea, presented a floral star to the Zionist chiefs with a little speech in Hebrew and Gladys Kone and Helen Fox, both 9 years old, presented a bouquet of roses to them. The crowds cheered wildly as Dr. Weizmann and Prof. Einstein stood up in the car to receive the flowers.

Solomon Elsner was toastmaster at the luncheon. Presenting the president of the world Zionist organization, he spoke of Dr. Weizmann's achievements as a chemist during world war. He told of Dr. Weizmann refusing a fortune which England offered as a gift for his services and of asking instead a hearing for the Jews.

Dr. Weizmann, in his speech, averred it was no sacrifice, but an honor and a privilege to be able to help the Jews realize the fulfillment of their dream of a homeland in Palestine. Till now the Jews waited for the world to answer their appeal—now the world is waiting for the Jews to respond to the opportunity which has been given them to redeem Palestine.

Dr. Weizmann expressed his great pleasure at being in Hartford where there are so many relatives of his or friends with whom he spent his childhood days in Mottele, Russia.

Ovation at Theater

Captain Abraham Tulin presided at the theater meeting. Here Dr. Weizmann and his associates received another big ovation. Dr. Weizmann said there were those who said the Zionists were impractical, and he knew it required cheerful, courageous people to meet the task of upbuilding Palestine. The song of toil and love is now heard among the pioneers, he said.

"I believe in the Jews," he said in closing his address. "I believe in the truth; strength and courage to you, work on and carry forward."

Dr. Schmarya Levine spoke poetically of the Jewish aspirations. England had given Palestine to the Jews, he said, because their right to it was indubitable. Palestine rebuilt by the Jews will repay the world by manifesting again its ancient genius. A liberated Palestine will help liberate the world, for out of Palestine will come again great ideas and great ideals. He said money for the upbuilding of Palestine was coming even from Poland and Ukrainia.

Professor Einstein said Palestine is a small land but its significance is greater than its physical size. How will it solve the Jewish problem? It will give the Jews a new civilization, a new culture and also renewed self-respect. Great responsibility calls forth self-respect and courage. The Zionists have accomplished the purpose of giving the Jews renewed courage and self-respect.

Reuben Taylor, president of the Zionist Regional union of Connecticut, said, pointing to the Zionist chiefs: "They come to us at a most critical period of Jewish life. Misery stalks among our brethren in Europe. The only light that shines for them is the light on the East."

The Hartford Daily Times, May 23, p. 2.

Hartford was a special destination for the Weizmann-Einstein mission. Around 1910, 80 percent of Hartford's Jewry was of East European origin: the Austro-Hungarian Empire, Romania, and Russia, which at the time included Latvia, Lithuania, Poland, and Ukraine. Isaac S. Hurevitz, whose name comes up several times in the pages of Hartford newspapers, was the first East European rabbi who settled in Hartford. The Agudas Achim Synagogue, also mentioned earlier, was the second East European orthodox congregation there.[81]

I hope that Einstein, in the car that triumphantly drove along Hartford's streets among waving flags and hands, realized that skipping the visit would have caused a scandal.

Baseball Makes U.S. Concentrate, Asserts Scientist Einstein

Famous Discoverer of Relativity Says Prohibition Shows Nation's Power— Movies Fine for Relaxation—His Theory Easy to Pure Thoughtists

Baseball unites the spirit of democracy, the "movies" are a grand thing and prohibition is good, Professor Albert Einstein, who recently startled the world with his theory of relativity, told a reporter for "The Courant" yesterday, but said he was not worrying about his theory for that reason declined to undertake an explanation of the theory.

Professor Einstein whose relativity theory has made the scientific world sit up and take notice and has aroused a desire among laymen and women for a non-technical explanation of what it is all about, is satisfied that persons who give their time to pure thought and to the scientific search for truth understand his theory and are not at all mystified. The distinguished Jewish scientist made this declaration when asked whether, as press reports stated, only a dozen persons understood his theory.

Shy of Reporters

The distinguished Jewish scientist was reluctant to speak about his theory, in fact at first flatly refused to say anything, but warmed up sufficiently to entertain some questions through an interpreter. Once he got started Professor Einstein was most pleasant and affable and went so far as to express his approval of the American "movies," said baseball was all right and even had a good word for prohibition. Hartford, he said, was a beautiful city.

Looks Like Music Master

Professor Einstein occupied a front row seat on the stage of the Capitol Theater yesterday. He is gray haired, medium sized and a shade over 40. He gazed at the thousands of faces in the auditorium with mingled pride and satisfaction, although he could not understand a word of what was being said in English and Yiddish. The pleasant faced man, whose long wavy hair gave him the appearance of a music master, listened eagerly to the words of the speakers who were appealing to the Jews of Hartford for support of the World Zionist Organization's plan to rebuild the Jewish homeland in Palestine.

World Zionist

The distinguish[ed] Jewish scientist is more interested in the rehabilitation of the Jewish homeland and the building of a Jewish university in Jerusalem than he is in discoursing indiscriminately on his theory of relativity, a scientific discovery the fame of which preceded his arrival in this country in company with Dr. Chaim Weizman, head of the World Zionist Organization, who was the head of Great Britain's chemical warfare bureau during the war.

Religion Not Involved

"Religion is not involved in my theory of relativity," said Professor Einstein, in response to a question as to the significance of his theory with that of Galileo who shocked the church with his discovery that the earth revolves around the sun.

"That is why my theory has not brought about so much general discussion," he added. He said that his theory did not answer the question as to the age of the world.

Touched by Suffering

While the appeal for funds was being driven home to the assemblage Prof. Einstein said: "They don't know what the suffering has been in Europe, they haven't suffered so much over here."

"This is a continuous rain of money. Hartford must be a rich city," he commented, as men, women and children poured their offerings into baskets held out to them by the ushers.

German Anti-Semitism

According to Professor Einstein, anti-semitism has grown stronger in Germany since the cessation of hostilities than it was before the war. Professors and students in the German universities, he said, were going to unheard of extremes in their hatred of the Jews, and in their treatment of Jewish students.

But things over here are different. Professor Einstein has evidently found his visit here gratifying and enjoyable. In the short time that he has been here he has seen much that has pleased him.

Movies, Baseball, Prohibition

"The movies? Oh, they are fine," said Professor Einstein. "I have seen many. They afford relief and relaxation." He was impressed with the magnitude of the industry in this country.

"What do you think of our national pastime? Have you seen baseball games?" he was asked.

"Baseball, baseball," repeated the scientist, "I've seen that game here. That's fine. It makes everybody concentrate on what is going on. It helps to unite the spirit of democracy."

"What do you think of prohibition?" the reporter asked Professor Einstein via the interpreter. "The ban on the drinking," added the interpreter.

Power of a Nation

"Oh," replied the professor smilingly, "that shows the power of a nation. And it's good."

"Is it true that the professor is an accomplished musician?" asked the reporter.

"No, no," was the ready response to the interpreter's question, which was the professor's only direct reply to the reporter. He admitted, however, that he found enjoyment in playing the violin.

Devoted to Hebrew University

Professor Einstein is greatly devoted to the idea of a Hebrew University in Jerusalem and said he was giving much of his time and efforts for the cause. He was especially pleased when Rabbi Abraham Novak informed him that the children of the Emmanuel Synagogue Sunday school had contributed $50 towards the university. The children had also donated $100 to the Zionist fund for Palestine restoration.

Protested Against War

Professor Einstein, who was born in Germany, lived in Switzerland during the war because he was one of the very few German professors who had the temerity to deny at the outbreak of the war, that Germany was not justified in making war.

The Hartford Daily Courant, May 23, pp. 1, 2.

Oh no! He did not deny it! By the way, he lived in Berlin during the war and only visited in Switzerland, where his divorced first wife lived with their two sons.

Zionist Delegates Wildly Acclaimed by Hartford Jews

Freedom of City Extended—$75,000 Raised in Free Will Contributions for Palestine Fund—Visitors Include Professor Einstein—Dinner, Speeches and Enthusiastic Meeting Follow Street Parade with 15,000 Spectators

Thousands of Hartford Jews yesterday paid high honor to world delegates of Zionism. The freedom of the city was extended and $75,000 was raised in free will contributions for the Palestine foundation fund. A street parade viewed by 15,000 spectators, with service men, who had fought in Palestine carrying flags, a dinner, with speeches by protagonists of the Zion cause, and an enthusiastic mass meeting, evinced the interest of Hartford Jews in the world wide movement. It was the day of days for the Jews. In the parade, at the dinner and repeatedly at the giant meeting which rocked the Capitol Theater with applause the old-time Jewish anthem "Hatikvoh" was sung with reverence and enthusiasm, its sad and plaintive music both composing and stirring the singers.

A telegram filed late Saturday night to the world delegates in New York city desired them to leave at Berlin the train to bring them to Connecticut, supplanting an earlier arrangement by which they were to leave at New Haven. The telegram failed to reach them, and when the reception committee awaited the delegation at the Berlin station it was learned that the visitors had followed the original schedule. Driving to New Haven the committee found the visitors at the Hotel Taft and took them by automobiles to the appointed rendezvous at Goodwin Park, where a long stream of escort machines was ready.

It was an hour after the announced time to start when the visitors were taken to Barnard Park, the formation point of the parade, but the delay had operated advantageously, for the genial summer sun had tempted thousands to the sidewalks.

Parade Is Different

It was a parade different from many held in this city. There was little of the pomp and circumstance of the military and much of deeper and abiding civic sentiment. With the flag of the United States the paraders carried the old and storied star of Zion and the tri-crossed banner of the empire which, by the Balfour declaration, held out the hope of restoration to the birthland of Jewry. Service men bore the flags. One was a lad in sailor blue. Two others were of the Jewish Legion which followed Allenby to Jerusalem and the Jordan. With them were a Camp Devens boy and a veteran of Vladivostok, these two as armed guards of the three flags.

A police platoon was preceded by motorcycle officers and followed by Colt's Band, the initial music being the "Second Regiment March." The service quintet came next.

Morris Tulin of No 54 Winthrop street, who served as a petty officer at the Naval Training School at Newport, was picked to bear the American flag. Samuel Charnowsky of No. 42 Bellevue street carried the Jewish flag and Jacob Robins of No. 22 Canton street the union jack. Both of them were in the Thirty-ninth Royal Fusiliers, having enlisted here, trained in Canada and gone by way of England, France, Italy and Egypt to Palestine. B. H. Teicher of No. 68 Capen street of a depot brigade at Camp Devens and Max Kovalsky of No. 304 Enfield street, who went in the Thirty-ninth Infantry's ordnance training detachment to Vladivostok, were the color guard.

Alderman William Raphael was marshal, occupying, with Louis Baron, Adolf Abrahamson, Abraham Fonstein, Miss Jeannette Fonstein and Miss Ruth Rabinovitz, the first car. The car following carried Dr. Chaim Weizmann and Professor Albert Einstein, the world Zionist leaders, with Captain A. Tulin. In the next Dr. Sch. Levin and Secretary N. Ginzburg of the delegation rode. The ten cars following carried the reception committee of which I. E. Goldberg is chairman. Near Keney Tower the column halted and flowers were presented to the two delegates. Continuing the route the column traversed High street to the Garde Hotel, where it disbanded.

Dinner was served by Caterer Max Walker at the Garde to about 150 men and women. Judge Solomon Elsner was toastmaster. After the singing of the Hatikvo, Dr. Weizmann spoke, telling of the transformation in Palestine and his observations there, which had led him to feel that with "just a little hard work" the Zionists would attain the dream of centuries. He said he was positive that Jews needed only to know the case, Palestine was developing. Money was the smallest thing needed. It was sure to come. Dr. Weizmann said he saw before him fine young fellows and he knew they would hear the call. It was not a sacrifice before them. It was a privilege. It was a great honor to help Zion. It was a challenge and a call. Dr. Levin spoke in Yiddish. He told about the pogrom. Turks and Arabs had killed Jews. Dr. Levin also told about the trouble in Jaffa.

Giant Mass Meeting Held

In the Capitol Theater, back of the stage a large star of Zion flag was flanked by two mammoth American flags and the British flag was also used. Hundreds in the audience, which filled the seats and the aisles, carried small American and Jewish flags. When the meeting opened children marched in from the wings, halted and faced in single rank toward the audience. The five service men entered, the orchestra played the opening bars of "The Star Spangled Banner" and the audience rose to sing this and then the Hatikvo. Alderman Raphael welcomed the guests in the name of the city, Mayor Brainard being unable to attend. He read the resolution passed by the court of common council as a testimonial to Dr. Weizmann. He added:

"Hartford takes pride in the fact that it has always recognized greatness and notable achievements. It is particularly honored today to receive and welcome such renowned leaders of the Jewish people who have come to our city in the interest of a great and noble cause.

"The restoration of Palestine to the Jewish people is an event for which the Jews have been praying and striving for many centuries. Now it is no longer a dream but a fact, a reality and you who are engaged so untiringly in the work of upbuilding and development of the Holy Land are entitled to the gratitude not only of the people of Israel but of the entire world—for we know and feel that the ancient prophecy is about to be fulfilled, and that Palestine is again to be a blessing to mankind."

Dr. Weizmann answered briefly and dwelt on the pride which the delegates took in the work, not alone because it was for the Jewish people, but also because it was for the cause of humanity. He said the delegates felt encouraged and happy in America, for they had found generous response, a wide sympathy for the cause. He thanked the alderman, the mayor and the city for the greeting.

A. Goldstein gave a greeting in Yiddish, pledging the wholehearted support of the Jews of Hartford. He assured Dr. Weizmann that what Dr. Herzl had begun they would finish.

Captain Tulin said that though he had been away from Hartford much of the time he was of Hartford in spirit. He told about inviting Dr. Weizmann four years ago. He called on Reuben Taylor, who said the time was not far past when one who argued for a Jewish nation was ridiculed, but today there was present a man who had worked untiringly and who had led along toward the Balfour declaration, favoring Palestine for the Jews. He praised the labor of Dr. Weizmann, the intellect of Dr. Einstein and the success of Dr. Levine. At mention of each the audience rose and cheered and waved hands and handkerchiefs in welcome. Mr. Taylor said this was a critical period in Jewish history. These delegates had come to America to tell of the new light that shone in the East. It was for American Jews to show that Palestine was as dear to them as to European Jews.

Alderman Raphael said that great moments in the history of Jews were distinguished by the appearance of men of zeal, vision, passion and willingness to sacrifice themselves, and who earned the name of prophets. Such a prophet was to speak, bringing a message. He was a man of whom no opponent could say but that his soul was pure, his disinterestedness complete. He was a man of the people, known by the people knowing them and having their confidence. Blessed was Israel for having men like Dr. Levin.

Dr. Levin spoke in Yiddish. He declared that the God of the Zionists was not a God of silver or copper or gold. While the Jews had been slaves they had scant opportunity to show genius. When they became free they produced genius. England and Germany has fought for the supremacy of the orient. Germany had gone down and Balfour and England had favored the Jews. It was possible for Zion to regain the old homeland.

Captain Tulin told the audience that the next speakers were to be heard not from the stage, but from the theater. He outlined the raising of a fund here to aid the cause. Captain Tulin and others on the stage aided in receiving and announcing the pledges and contributions which came in volume from the body of the house. Ushers circulated along the aisles and a girl with an adding machine on the stage sought to keep even with the increasing flow of checks, cards and cash across the footlights. But she could not, such was their volume. Effort will be made to classify and arrange the cards today.

Many Gifts Subscribed

Among the gifts were the following: Joseph Silver, $2,000; Nirenstein & Schwartz, $2,000; Hadassah Chapter, $2,000; Tikvoh Camp, $2,000; Koretzer Congregation, $2,000; Vincent Levy, $2,500; Hyman Kaplan, $1,500; Tulin, Toft & Tulin, $1,500; Morris Cohn of New Britain, $1,000; Mr. Bassevitch, $1,000; Israel Silver, $1,000; L. E. Goldberg, $1,000 and $1,000 more conditional on Hartford Jews raising a total of $50,000; Herman P. Kopplemann, $500; David A. Rosso, $500; Walter Beckman, $500; H. Eisenberg, $500; F. Finkelstein, $500; William Rulnick, $500; Mr. and Mrs. Libman, $500; Louis White, $500; Solomon Tulin, $500; Jacob Sneker, $500; Young Men's Hebrew Association, $500; Young Women's Hebrew Association, $500; Israel Sherman, $500; Ralph Kolodney, $500; Abe Wolsky, $500; Jacob Sowolsky, $500; H. Freeman, $250; Judge Solomon Elsner, $250; Samuel Tulin, $250; B. Rappaport, $250; Leviat S. Knoek, $250; Isaac Paul, $250; Meyer Greenberg, $250; J. Selitzky, $250; Sisterhood of the Emanuel Synagogue, $250; Morris Taylor, $250.

B. Wachtel, $300; Harry Roth $300; William Vogel, $300; David Traub, $300; Jacob Richman, $200; King's Cigar Company, $200; M. H. Silverman, $200; Israel Garber, $200; David Freedman, $200; Max D. Berman, $200; H. Shimmelman, $200; M. J. Dubin, $200; F. H. Levy, $200; M. C. Goldberg, $200; H. Freeman, $200; H. Epstein, $100; Morris Apter, $100; J. Rosenbaum, $100 (cash); Charles Siegel, $100; Mrs. I. Cohen, $100; Reuben Norwitz, $100; Alex. Tulin, $100; Mrs. K. Goldberg, $100; Isaac L. Cohen, $100; A. Levin, $100; I. L. Selvin of Bristol, $100; S. Levy, $100; N. Rosenfeld, $100; A. Brodsky, $100; M. Snyder of New Britain, $100; F. L. Simmons, $100; Mrs. F. H. Levy, $100; Israel Nivitz, $100; Samuel E. Herrup, $100; R. Rubin, $100; A. Shapiro, $100; M. Larrabes of South Manchester, $100; W. Y. Barbara, $50 (for the university); R. Elkin $50; M. Ginsberg, $50; S. D. Tulin, $50; Jacob Manning, $50.

It was announced that Ellington Jews had pledged $1,000 and the Jews of Norwich $20,000. A woman whose name was not given contributed a box of jewelry. Sam Cohn announced that ex-Mayor Kinsella and Dr. J. H. Naylor had given $50. Good natured rivalry was shown when an announcement that the Y.M.H.A. had subscribed $500, $100 to be paid annually for five years, was followed instantly by an announcement made by a woman that the Y.W.H.A. was going to do the same thing.

During a brief lull it was announced that, feeling that the meeting would surmount the top, it was arranged the night before to have a certified check ready to hand over to the delegates. A number of local workers had underwritten to contributions in advance to the amount of $25,000 and the check was delivered to Dr. Weizmann. Among those underwriting were S. D. Tulin, I. E. Goldberg, L. B. Toft, M. M. Taylor, G. B. Swartz, Samuel Hirschman, E. Nirenstein, Hyman Kaplan, F. Finkelstein, Reuben Taylor, Joseph S. Silver, David Kaplan and George Levin.

Refers to Balfour

Captain Tulin told of hearing Arthur J. Balfour, once prime minister of Great Britain, refer to talking in 1906 with a young Jew in Manchester, England, about Zionism. The captain was in Albert Hall in London last summer and told about the power Dr. Weizmann had over a gathering of 15,000 people. Dr. Weizmann was called on again at yesterday's meeting and told anew of the place of Zion.

Professor Albert Einstein was the final speaker, using German, his only tongue. He said he felt that the future of Judaism was tied up with the rehabilitation of the homeland. He said he knew the movement would succeed.

When the meeting concluded at 5:30 o'clock the delegates started for New Haven, where they spoke in the evening.

The Hartford Daily Courant, May 23, pp. 1, 2.

Two days earlier New Haven gave the delegates the freedom of the city (see below).

Camp Devens near Boston was a huge military training center up to 1996.

Freedom of City for Noted Jews

The Board of Aldermen at the special meeting last night adopted a resolution granting a "welcome and freedom of the city" to Prof. Albert Einstein, famous propounder of the theory of relatively [*sic!*] and Prof. Chaim Weizman who will be here tomorrow in the interest of the Palestine fund.

The resolution was presented by Alderman Peter J. McNerney, who stated that New Haven should extend this honor to the distinguished visitors as is being done by other cities in the country.

The New Haven Journal-Courier, May 21, p. 3.

$75,000 Raised by Local Jews in Single Night

Welcome the World Zionist Commission with Huge Contribution—Einstein's Visit Is Brief—Leaves for New York after Short Chat with Committee Here

First-hand information of the present situation in Palestine was received here last night by about 5,000 local Jews who turned out to welcome the World Zionist Commission at a mass meeting in the Arena. In connection with the visit of the commission, it is estimated that $75,000 was raised in contributions and pledges made at the meeting and the banquet preceding the meeting. Most of the individual contributions were about $1,000.

In the party of distinguished visitors who came here at 7 o'clock last night by automobile from Hartford were Prof. Chaim Weizman, president of the World Zionist organization; Prof. Albert Einstein, noted scientist; Dr. Schmarya Levin, noted writer on philosophy, and several secretaries. A large delegation, in automobiles, accompanied the party from Hartford as far as Berlin, and a similar delegation met the party on the outskirts of this city.

Einstein's Visit Brief

Prof. Einstein, discoverer of the theory of relativity, after chatting a few minutes with members of the reception committee, Mayor FitzGerald and a delegation from the faculty of Yale university headed by Prof. Irving Fisher, left immediately for New York to keep an engagement for the evening.

A banquet was given for the visitors in Fraternity hall (formerly Harmonie hall) in Elm street by a committee of prominent Jews of this city at which the speakers were Prof. Weizman, Rabbi Alexander Levine and Mayor FitzGerald. The members of the commission then were escorted to the Arena where, meanwhile, the people who had come to the meeting, scheduled to take place at 7:30, had been waiting about two hours. A band had been dispatched from Harmonie hall when it became known that the visitors, who arrived late, would not get to the Arena on time, and the impatience was somewhat relieved.

Prof. Weizman, in his address at the Arena, told the people, in his native tongue of the administration of the Keren Hayesod (Palestine foundation fund) through the instrumentality of which irrigation, immigration, electrification and other public and semi-public work is being done in Palestine. He told how the Zionist organization has acquired enormous tracts of land and is now acquiring still more land. There are plenty of Chalauzim (Jewish volunteers), he said, mostly university students, in all the European countries who are willing and anxious to enter Palestine and if need be dig ditches and build roads or do any other work which might help to re-build Palestine.

But in order to enable these Jewish volunteers to emigrate to Palestine, he said, and to make preparations to receive them, it behooves the Jews of America to contribute as much money as they possibly can. He compared the sacrifices of the Jewish volunteers entering Palestine who are giving their all and their lives if need be, with the small part that American Jews are being asked to do in the building of Palestine by their financial contributions.

Prof. Einstein, before he left, also discussed with prominent local Jews his plans for the building of a university in Palestine.

Dr. Schmarge Levin and Rabbi J. Lovenberg also spoke at the meeting in the Arena.

Weizman Meets Kin

One of the interesting incidents of the welcome to the commission here was the meeting of Prof. Weizman with local relatives of his whom he recognized, though seeing them for the first time in 25 years.

The following were in charge of arrangements:

Samuel Nathanson, chairman; Samuel Eskin, director; Louis Sachs and Pinkus Churgin, secretaries; and David Steinberg, treasurer.

Reception committee—Attorney Charles Cohen, Justice Commissioner

Hyman Jacobs, Rabbi J. Levenberg, Rabbi Alexander Levine, I. Kauffman and Judge Jacob Caplan.

Speakers committee—Rabbi Levenberg, Rabbi Levine, Attorney S. J. Nathanson, Attorney Charles Cohen, Dr. Bercinsky and Samuel Silverman.

Music committee—Samuel Eskin, D. Steinberg and I. Kauffman.

Ladies reception committee—Mrs. S. Eskin, chairlady; Mrs. S. Caplan, Mrs. A. Komroff, Mrs. N. Levine, Mrs. H. White, Mrs. A. Shultz and Mrs. A. Lander.

Prof. Weizman and the other members of the Zionist commission will go from here today in Worcester, Mass., where they will attend a meeting tonight.

Prof. Einstein and Prof. Weizman were given an official welcome and the freedom of the city by the Board of Aldermen at its special meeting last Friday night.

The New Haven Journal-Courier, May 23, pp. 1, 2.

> Einstein had been invited to deliver a course of Silliman Lectures (a series of eight to twelve lectures) at Yale early in 1921 "to form the basis of a volume which would have been published at the expense of the university," but at the time he did not accept the invitation because it would have been only one stop on a longer lecture tour in the United States, which ultimately came to grief.[82]
>
> A second invitation was presented by Professor Irving Fisher when Einstein was already in New York. Fisher proudly mentioned that he had studied mathematics in Berlin and had heard a few lectures of Helmholtz. He had tried to contact Einstein in person on April 9 and invite him to Yale, but he had failed to meet him. Einstein gave a polite but indecisive answer, referring to his many obligations.[83]
>
> One of Yale's most prestigious persons, Bertram Borden Boltwood, professor of radiochemistry, was not very enthusiastic about Einstein the Zionist: "Thank heaven, Yale did not give Einstein a degree. We escaped that by a narrow margin. If he had been over here as a scientist and not as a Zionist it would have been entirely appropriate, but under the circumstances I think it would have been a mistake."[84]

17

Professor Einstein and the Hat

✦

Monday, May 23

The Adventure of Professor Einstein and the Hat

Scientist Fails to Apply Relativity Theory

Prof. Albert Einstein can estimate how much the light of a distant planet is pulled out of its course by the attraction of the sun, but there is nothing in the theory of relativity that tells him how to map the erratic wanderings of a runaway hat. And even if he is so learned that there are only twelve men in the world who can understand his book, he doesn't yet know that the easiest way to recover a lost headpiece is to let someone else chase it.

It was because Professor Einstein has not learned these things that a Harlem crowd on a recent date was treated to the sight of a world-famous scientist frantically dodging in and out along motor cars and trucks on crowded upper Seventh Avenue in pursuit of what an American would have called his "lid".

That day professor Einstein had accepted an invitation to be the guest of the Faculty Club of New York University and plant a tree on the campus at University Heights.

Prof. Thomas W. Edmonson, Prof. John C. Hubbard and Captain Henry C. Hathaway, director of public occasions for the University, charged with the duty of escorting the guest to University Heights, decided that Captain Hathaway's open car would be just the *thing*. When the party had reached the neighborhood of 133rd St. a wild cry from the back seat caused Capt. Hathaway to throw on the emergency break. He turned his head in time to see Professor Einstein make a flying leap out at one door of the car, his companions emerging only a moment later through the other.

Einstein and President Ernest Elmery Brown at New York University. They planted a sapling
in front of the area that now is the cobblestone frontwalk to Tech II.
Courtesy New York University.

Professor Einstein affects a hat of the wide rimmed variety and when a blast of wind snatched it from its owner's head it "lit running" and was picking a speedy course approximately down the center of Seventh Avenue.

By the time he could bring his car to a stop at the curb, Captain Hathaway was hopelessly distanced, the chase already having gone far, with the champion of relativity easily leading his two fellow passengers. Professor Edmonson expended a quantity of much-needed breath in shouting for somebody to stop that hat, but after all, it was a sewer catch basin that brought the case to an end, and it was Professor Einstein who was in at the death, smiling and triumphant over the discomfiture of his less speedy escort.

—New York Sun.

The Campus, June 4, p. 5.

Einstein was asked twice for a visit and lecture at New York University on April 5 and April 22. He assented, then withdrew his promise because of a "most important affair in connection with my mission." He accepted a new invitation of May 9, however, and, as you have just read, he managed to demonstrate his athletic abilities too.[85]

18

To Be in Cleveland

✦

Friday, April 22

Mizrachi Zionists to Honor Zionist Leaders

The Mizrachi Zionist Organization of Cleveland will hold a Passover celebration and entertainment in the auditorium of the new Talmud Torah,
E. 55th street, Wednesday evening, April 27. The meeting will celebrate the
coming of Dr. Chaim Weizmann, Menachem Mendel Ussishkin and Professor Albert Einstein as a Zionist Commission to this country. Plans will
be formulated for the initiation of the Keren Hayesod campaign in this
city. An elaborate program has been arranged. Cantor Albert Levin and
the Talmud Torah Choir will sing Hebrew folk songs and several artists of
great repute will also render vocal and instrumental selections. Rabbi Samuel Benjamin of the Cleveland Jewish Center, Isaac Carmel, noted Zionist
leader, Rabbi Braver of Akron and Rabbi Pelkovitz of Canton will speak. No
admission fee will be charged and the public is invited to attend.

The Jewish Independent, April 22, p. 5.

Friday, April 29

Visiting Zionist Leaders to Be in Cleveland on May 25

**President Chaim Weizmann of World Zionist Organization and Prof. Einstein
in Party—Clevelanders Arrange Reception—Executive Committee of Cleveland Zionist District Urges That Negotiations Be Resumed by Dr. Weizmann
and Leaders of Zionist Organization of America**

Word has been received that the Zionist Commission to the United States, including President Chaim Weizmann of the World Zionist Organization, Menachem M. Ussishkin and Benzion Mossinsohn, will be in Cleveland May 25. Dr. Albert Einstein, world-famed discoverer of the theory of relativity, will be with the commission. Dr. Einstein is appealing for support for the Hebrew University in Jerusalem. Mrs. Einstein and Mrs. Weizmann will also be in Cleveland.

[The paragraphs on a conference of Orthodox congregations supporting the Keren Hayesod are omitted, as are those on a resolution of the Cleveland Zionist District on refraining from participation "in any and all campaigns for funds" until negotiations are resumed between the World Zionist Organization and the Zionist Organization of America.]

The committee to organize the Weizmann reception in Cleveland decided to issue a call for a city-wide conference to be held May 8 and to invite Louis Lipsky of New York as the main speaker. The committee unanimously decided to declare itself as the arrangements committee for the establishment of a permanent Keren Hayessod bureau in Cleveland. The members of the committee are: Dr. M. Garber, chairman; Isaac Carmel, secretary; I. Feigenbaum, treasurer; Rabbi Samuel Benjamin, Rabbi Moses Schussheim, Jonas Gross, Aaron Garber, S. Rocker, Leo Weidenthal, W. R. Widenthal, M. Savin, Henry Feigenbaum, Sam Garber, Isidor Rothstein, Nathan Tempkin, Abe Lev, M. Herwald, Dr. Burstein, M. Blitstein, M. Kamet, Samuel Ismach, Mrs. M. Garber, Mrs. M. Schussheim and Mrs. Isidor Rothstein.

[Weizmann's appeal on behalf of the Keren Hayesod, part of which is quoted in the article, is omitted.]

A ladies' committee was organized at a meeting held Monday evening, April 25, at the Home of Mrs. Isidor Rothstein, 9104 Parkgate avenue. Among those present were Mrs. M. Schussheim, Mrs. Milton Weil, Miss Lottie Bialosky, Mrs. L. E. Lashman, Mrs. Mandelkorn, Mrs. M. Garber, Mrs. J. Flock, Mrs. A. Krohn, Mrs. L. W. Klusner, Mrs. I. Belkowsky, Mrs. H. E. Eisler, Mrs. H. Tabakin, Miss Fell, and Miss Kramer.

A committee consisting of Mrs. I. Rothstein, chairman; Mrs. Milton Weil, treasurer; Misses Kramer and Fell, secretaries; executive committee, Mrs. M. Garber, Mrs. Klusner, Mrs. Schussheim, Miss Bialosky, Mrs. Lashman and Mrs. Mandelkorn was appointed to arrange a special reception for Mrs. Weizmann and Mrs. Einstein, when they visit Cleveland. The meeting also decided to launch a "Jewelry campaign," which is part of the

Keren Hayessod, and which was originally started by Mrs. Weizmann, and of which Lady Herbert Samuel is chairman.

The Jewish Independent, April 29, pp. 1, 6.

Friday, May 6

To Confer on Weizmann Reception

Meeting Called for Sunday Afternoon by Representatives of Jewish Organizations—Louis Lipsky Will Address Gathering—Mayor FitzGerald to Name Committee of 100

For the purpose of organizing a reception in honor of the Zionist Commission to the United States, a conference of representatives of Jewish organizations has been called for Sunday afternoon, May 8, at 2 p.m., at the Cleveland Jewish Center, on E. 105th street. The Commission will be in Cleveland on Wednesday, May 25.

Louis Lipsky of New York will address the meeting on Sunday. He will deal with the present situation in the Zionist movement. A special committee will visit this Saturday morning all synagogs and appeal for their participation in the mass conference.

Mayor W. S. FitzGerald will name a citizens' committee of 100 to have charge of the official welcome to the distinguished guests. Among the visitors to be received by the city are President Chaim Weizmann of the World Zionist Organization, Dr. Albert Einstein, discoverer of the theory of relatively [sic], Menachim M. Ussishkin, and Benzion Mossinsohn. Mrs. Einstein and Mrs. Weizmann will also be in the party.

The meeting for next Sunday has been called by the committee that is planning arrangements for the Weizmann reception.

[I omit here Weizmann's open letter to Mack on an issue of the ongoing controversy between them.]

The Jewish Independent, May 6, p. 1.

Sunday, May 8

Arrange for Reception in Honor of Zionist Commission

Committees to Take Charge of Events on Occasion of Visit of Distinguished Scientists—Mayor Names Committee of 100—Mass Conference at Jewish Center Votes to Support World Zionist Organization—Approves Keren Hayesod in Formal Resolutions and Decides to Organize a Branch Bureau in Cleveland

Seven hundred delegates, representing over one hundred Jewish organizations, participated in a mass conference Sunday afternoon at the Cleveland Jewish Center, E105th street and Grantwood avenue, at which the coming visit of the Zionist Commission to the United States was discussed. The conference expressed its gratification at the opportunity to greet the three eminent visitors, Dr. Chaim Weizmann, Prof. Albert Einstein and M. M. Ussichkin, who are due in Cleveland on May 25.

The conference took up the question of the "Keren Hayesod" as advocated by the World Zionist Organization, and unanimously decided to support the World Organization whose leader and spokesman Dr. Weizmann, has authorized the committee of the Weizmann, Einstein, Ussichkin reception, to open a bureau for this fund in Cleveland.

The conference was opened by Dr. M. Garber, who explained the purpose of the meeting. Isaac Carmel, secretary, tendered a report of the activities of the temporary committee and made a short and touching address on the victims who fell in the Jaffa riots. The audience rose and stood for one minute in silence in honor of those who fell as martyrs and heroes for their country and people.

At the suggestion of Aaron Garber, the conference unanimously decided to organize a bureau in Cleveland for the collection of funds for the "Keren Hayesod" as a branch of the bureau opened in New York by the World Zionist leaders.

Rabbi Samuel Benjamin moved that this assembly organize itself as a body to arrange a proper welcome for Dr. Weizmann, Dr. Einstein and M. Ussichkin and it unanimously carried. Dr. M. Garber was elected chairman; Isaac Carmel, secretary; Aaron Garber, Max Cohen and S. Ricker, vice chairmen.

A resolution committee elected at the conference, consisting of Sol Hurvitz, S. Epstein, I. Feigenbaum, Rev A. Levine, D. Gara, Rabbi S. Benjamin,

Rabbi M. Schussheim, Mrs. Jennie Zwick, I. Applebaum, J. Handelman, I. Geffen, H. Pollack and S. Lurzman offered resolutions defining in detail the sentiments of the conference as already expressed in the resolutions sponsored by A. Garber and Rabbi Benjamin. The resolutions follow:

"Resolved, that we, the Jews of Cleveland, assembled in conference on the 8th day of May 1921, hereby re-affirm our loyalty to the World Zionist Organization and its leaders and declare that the Keren Hayesod as established and formulated by the said World Zionist Organization is the most feasible method of raising funds in the United States for the upbuilding of Palestine as a Jewish homeland, and we pledge ourselves to do our utmost for the success of this fund."

Ab Goldberg, famous Hebrew writer and member of the National Executive Committee of the American Zionist Organization, addressed the conference and reviewed the efforts of Dr. Weizmann, which preceded the Balfour Declaration, the San Remo Decision, the appointment of Sir Herbert Samuel as high commissioner in Palestine, and the recognition of the World Zionist Organization in the mandate as the Jewish agency to build up the Jewish homeland.

Louis Lipsky, Zionist leader, received an ovation when he rose to address the conference. Mr. Lipsky strongly criticised leaders of the American Zionist Organization for their attitude on the decisions of the World Conference in London. He demanded that Judge Mack and his friends either resign or follow the leadership of Dr. Weizmann.

The following were named as members of the Weizmann, Einstein, Ussichkin reception committee: M. D. Shanman, Max Cohen, J. K. Arnold, I. Feigenbaum, Morris Speer, Dr. J. Flock, Dr. I. Gittleson, A. Lipskovitz, Abraham Sacks, M. Savlin, Sol Horvitz, Jonas Gross, Dr. M. Garber, Aaron Garber, Isaac Carmel, Rabbi Samuel Benjamin, Rabbi Moses Schussheim, Mrs. M. Garber, S. Ricker, Louis Cohn, Louis Katz, D. Gara, Mrs. Jennie Zwick, Lewis Drucker, Sam Caplan, Max Salzman, Mrs. S. P. Burstein, Mrs. Volk, A. Lerzman, Mrs. S. P. Burnstein, W. R. Weidenthal, Dr. I. Zwick, Z. Kalinsky, Dr. S. P. Burstein, M. Hervald, Rev. Botivin, W. Kamin, M. Blitstein, S. Yirmach, Sol. Epstein, Dr. Milcoff, S. Resnick, M. Gandel, M. Palis, I. Finkelstein, David Eidelberg, G. Laufman, S. Schorr, M. Goldstein, L. Kamin, P. Lamden, M. Abrahamson, A. Chertoff, Mrs. F. Dolinsky, Mrs. M. Schussheim, Mrs. August, Mrs. G. Klein, J. Neshkes, Joseph Feder, Mrs. G. Laufman, Mrs. I. Rothstein, Mrs. Max Cohen, Miss Sarah Allen and Mrs. I. Timen.

Announcement was made at the meeting of the appointment of a reception committee to extend an official greeting to the guests in behalf of the city. Mayor W. S. FitzGerald is honorary chairman of this committee and W. R. Weidenthal chairman. Other members are J. K. Arnold, Joseph E. Arnoff, Mrs. I. Altfeld, E. M. Baker, Newton D. Baker, Rabbi Samuel Benjamin, Alex Bernstein, S. J. Bialosky, Mrs. A. T. Brewer, Marx Berne, Miss Lottie Bialosky, Rev. Dan F. Bradley, A. A. Benesch, Alexander C. Brown, Mrs. S. P. Burstein, G. S. Black, Dr. Charles F. Brush, R. C. Bulkley, Herman C. Baehr, Judge Maurice Bernon, Isaac Carmel, Louis Cohn, Thomas C. Cagwin, Max Cohen, Thomas Coughlin, Prof. Charles W. Coulter, Sam Caplin, Max Efros, Mayor W. S. FitzGerald, Paul L. Feiss, I. Feigenbaum, Judah Fineberg, Mrs. Julius Fryer, Thomas S. Farrell, Herman H. Finkle, Dr. J. Flock, Max Friedman, Morris Friedman, Rabbi Solomon Goldman, James R. Garfield, C. A. Grasselli, Mrs. Moses G. Gries, D. Gara, Dr. M. Garber, Aaron Garber, Sam Garber, Jonas Gross, F. H. Goff, Dr. R. Gittleson, M. Graff, Mrs. M. Garber, Edward L. Harris, Rev. Joel Hayden, Munson Havens, Dr. Charles S. Howe, Eric C. Hopwood, Mrs. Siegmund Herzog, Paul L. Hirsch, Robert Hoffman, Miss Esther Icove, Al. N. Jappe, Rabbi Leo Jung, Judge Thomas M. Kennedy, Benjamin Karr, Siegmund Korach, Louis Katz, Benjamin Lowenstein, Nathan Loesser, Daniel Lothman, Mrs. Edward Lashman, Judge Manuel Levine, Dr. A. S. Maschke, Dr. D. C. Miller, Frank H. Minor, Mrs. M. A. Marks, Rev. Francis T. Moran, J. Makoff, Maurice Maschke, M. J. Mandelbaum, Edward A. Moss, Max E. Meisel, Dr. John Newberger, J. C. Newman, J. R. Nutt, Max M. Ozersky, Miss Effie K. Oppenheimer, John T. Owens, J. B. Parker, Henry Rocker, Isadore Rothstein, S. Rocker, Sig. Ravinson, W. H. Rose, Rabbi A. H. Silver, Rabbi M. Schussheim, Judge Samuel H. Silbert, M. D. Shanman, Harry Simon, Henry Spira, Morris Speer, Jacob Stacel, Alfred L. Steuer, David Seidman, Rev. Dr. D. T. Scullen, Samuel B. Tilles, Maxwell F. Tielke, S. Ulmer, Judge Willis Vickery, Rabbi Louis Wolsey, Leo Weidenthal, W. R. Weidenthal, Dan Wertheimer, Nathan Zweig, Mrs. Jennie Zwick.

The Jewish Independent, May 13, pp. 1, 5.

The Balfour Declaration (see p. 39) was reaffirmed at a conference of the Allied Supreme Council held at San Remo, April 19–26.

As used above, a "mandate" means a sphere of interest in a territory that had been part of the Turkish Empire before it was defeated in World War I.

On May 1, a fight arose in Jaffa between Jewish Bolsheviks and Socialists. The Arabs who witnessed the incident started ransacking Jewish stores and homes; Arab policemen joined them. The riots continued for a week, spreading to other Jewish settlements as well. In all, 47 Jews and 48 Arabs were killed, 146 Jews and 73 Arabs wounded.

Friday, May 20

World Zionist Leaders Will Be Welcomed at City Hall

Banquet at Jewish Center and Auto Parade Included in Reception—Mass Meeting at Masonic Hall—Dr. Weizmann Will Be Speaker at Gathering—Mayor FitzGerald to Deliver Address of Welcome—Delegation from Other Ohio Cities to Attend Mass Meeting

The Zionist Commission to the United States which is due in Cleveland this coming Wednesday, May 25, will be welcomed at City Hall upon arrival. The Weizmann Reception Committee, which represent over 100 Jewish organizations, has arranged the program.

At 4:30 Wednesday afternoon there will be an auto parade which will start from the Hollenden Hotel where the guests will have their headquarters. The parade will pass through Woodland avenue and will stop for ten minutes at the Hebrew School and Institute at 2491 E. 55th street, where the Zionist leader will be welcomed by the staff and pupils of the Hebrew School.

The parade will disband at the Cleveland Jewish Center, E. 105th street.

A banquet for 600 people will be served at the auditorium of the B'nai B'rith Building in honor of Prof. and Mrs. Weizmann, Prof. and Mrs. Einstein, M. M. Ussischkin and Dr. Schmariah Levin.

The event of the day will be the mass meeting which is to be held at the auditorium of the new Masonic Hall, to be addressed by Dr. Weizmann, M. M. Ussischkin, Dr. Schmariah Levin, Mayor FitzGerald, Governor Davis and many other prominent citizens. Among those who have accepted invitations to be present at the new Masonic Hall are ex-Secretary of War, Newton D. Baker and Dr. Charles S. Howe, president of the Case School of Applied Science.

On Thursday morning, May 26, at 10 a.m., a Regional Conference of Zionist Districts, Mizrachi Societies, Polei Zion Societies, Zeirei Zion Groups, Haddassah Chapters, Senior Young Judea Clubs will be held at the assembly rooms of the Hollenden Hotel.

Dr. Weizmann will address the conference which takes up the question of arranging the Keren Hayesod Bureau in the State of Ohio.

Cleveland Jewish women are arranging a special noon luncheon at the Statler Hotel in honor of the Zionist Commission.

The committee of seventy-five, which was elected at the conference of Jewish organizations will meet Saturday evening, May 21, at the Hebrew School 2491 E. 55th street, for the purpose of making final arrangements of the visit of the Zionist leaders. The city will be decorated with American and Zionist flags in honor of the occasion.

Isaac Carmel, secretary of the Weizmann reception committee, has received telegrams from all cities of Ohio that they will have their delegations in Cleveland May 25 to meet Dr. Weizmann.

The Jewish Independent, May 20, p. 1.

Will Honor Distinguished Guests

Twenty-two Hundred Jewish Women of Cleveland to Attend Reception and Luncheon in Masonic Hall, Thursday, in Honor of Mrs. Chaim Weizmann and Mrs. Albert Einstein

Final plans have been announced for the reception which Cleveland Jewish women will tender Mrs. Chaim Weizmann and Mrs. Albert Einstein, when the wives of the noted scientists and Zionist leaders in company with their husbands, visit Cleveland next Wednesday and Thursday. Mrs. J. K. Zwick is at the head of the organization making the arrangements. The reception is to be the occasion at which Mrs. Weizmann will be presented with a fund raised by the Jewish women of Cleveland by the contribution of old jewelry to be devoted to the interest of Palestinian settlers.

A luncheon, next Thursday, in Masonic Hall, 3515 Euclid avenue, to be attended by approximately 2,200 women, is to be the chief event in the reception. Preparations for this luncheon are being made by an arrangement committee that embraces the leaders of practically every circle of Jewish women's activities in Cleveland.

Among the members of this committee are Mrs. S. Herzog, Mrs. Julius Fryer, Mrs. Sol. R. Bing, Mrs. Marcus Feder, Mrs. Charles Rosenblatt, Mrs. Louis Rich, Mrs. I. Altfeld, Mrs. S. P. Burstein, Mrs. H. Frankel, Mrs. D. Gara, Mrs. C. Ginsburg, Mrs. A. Mendelsohn, Mrs. S. Sampliner, Miss Tillie Cohen, Mrs. Charles Jaskalek, Mrs. I. Rosenstein, Mrs. J. Janowitz, Mrs. L. Wolk, Mrs. William P. Yelsky, Mrs. L. E. Lashman, Mrs. L. W. Klusner, Mrs. I. Bloch, Mrs. M. Garber, Mrs. S. L. Levy, Mrs. Max Cohen, Mrs. M. Mendelke, Mrs. J. Newman, Miss Lottie Bialosky, Mrs. Sol Friedman and Mrs. J. K. Zwick.

Co-operating with the Mayor's Reception Committee that has been appointed to receive Professors Weizmann and Einstein will be a Women's Auto Corps consisting of automobiles driven by women entirely and carrying the members of Women's Reception Committee of One Hundred. This reception committee is headed by Mrs. George Kline and Mrs. S. Sampliner, and its members, together with the members of the Luncheon Arrangement Committee, will serve as hostesses at the banquet on Thursday. Mrs. Julius Fryer has been named chairman of the hostess committee.

A special Decoration Committee has been appointed to arrange for the appropriate decoration of the automobiles and the hall in which the luncheon is to be held. This committee is headed by Sol Friedman and Mrs. A. Mendelsohn.

The chief event at the luncheon will be the presentation to Mrs. Weizmann of a treasure chest containing the jewelry which the Cleveland women have contributed for the Palestinian settlers. A special committee handling the jewelry collection has been named with Miss Tillie Cohen and Mrs. Samuel Volk-heading it.

Much beautiful jewelry has already been contributed, according to the leaders of the campaign. Many women have been donating the entire or greater parts of their family plate, while others have given all their personal jewelry. Stations at which jewelry may be brought are the B'nai B'rith Building, 7103 Euclid avenue, where Mrs. J. Newman is in charge; the Jewish Center, Hampden avenue and E. 105th street, with Mrs. L. Wolk in charge; and the new Talmud Torah Building, with Mrs. S. P. Burstein.

Reservation for the luncheon by women wishing to attend may be made by applying directly to Mrs. L. E. Lashman, 5000 Woodland avenue, chairman of the Reservations Committee; Mrs. Henry Frankel, and Mrs. L. Levy, who are the other members of the committee, or any of the other women who are participating in the jewelry drive.

The final general meeting of the women participating in the drive, and of those making the various arrangements, will be held Monday morning at 10 o'clock in the B'nai B'rith Building.

In addition to those who are on the Committee on Arrangements, the following women are included in the reception committee: Mrs. Ida Arnold, Mrs. H. August, Mrs. H. H. Abramson, Mrs. Morris Abrams, Mrs. L. Abrams, Mrs. Charles Brown, Mrs. Charles Arnson, Mrs. Alex Bernstein, Mrs. L. Auerbach, Mrs. A. Auerbach, Mrs. H. Bassett, Mrs. M. Berkovitz, Mrs. H. Berkovitz, Mrs. S. Brenner, Mrs. Joseph Berman, Mrs. H. Baskind, Mrs. E. D. Benedict, Mrs. J. L. Bubis, Mrs. Blachman, Mrs. M. A. Black, Mrs. Henry Dettlebach, Mrs. H. Dolinsky, Mrs. F. Dolinsky, Mrs. H. W. Dairs, Mrs. Joseph Buchman, Mrs. Ned Cohen, Mrs. B. F. Corday, Mrs. David Eaffy, Mrs. M. Eichorn, Mrs. J. W. Epstein, Mrs. J. Chertoff, Mrs. E. A. Cohen, Mrs. J. Flock, Mrs. S. J. Frankel, Mrs. S. Forstein, Mrs. George Heinz, Mrs. L. Barons, Mrs. E. Halle, Mrs. M. Halle, Mrs. T. Fishel, Mrs. S. J. Farber, Mrs. J. H. Goodman, Mrs. A. L. Goldman, Mrs. Joe Harris, Mrs. H. I. Jacobs, Mrs. H. M. Krause, Mrs. Louis Kline, Mrs. C. Kirtz, Mrs. J. Krivosky, Mrs. H. Greenbaum, Mrs. M. Kahn, Mrs. H. S. Kollin, Mrs. D. Katz, Mrs. H. M. Lewis, Mrs. J. M. Lewis, Mrs. M. Sacheroff, Mrs. L. Mitchell, Mrs. William E. Rothstein, Mrs. M. Rothman, Mrs. Hollander, Mrs. K. Sarefsky, Mrs. A. B. Rippner, Mrs. D. Shaw, Mrs. B. Siegel, Mrs. K. Kolinsky, Mrs. S. Schwartz, Mrs. B. Sands, Mrs. A. A. Katz, Mrs. Rousuck, Mrs. J. Katz, Mrs. Wien, Mrs. Elias, Mrs. J. Joseph, Mrs. H. Moss, Mrs. N. Loeser, Mrs. L. Stieglitz, Mrs. A. Lehman, Mrs. L. Wolin, Miss Ida Schott, Mrs. Bessie Spanner, Mrs. Manuel Halle, Mrs. Manuel Levine, Mrs. Sam Hartman, Mrs. I. Gogolich, Mrs. S. Silbert, Mrs. J. Marks, Mrs. J. Rose, Mrs. C. Katz, Mrs. Guggenheim, Mrs. H. Grassgreen, Mrs. J. Laufman, Mrs. Geo. Kline, Mrs. A. Tabakin, Mrs. A. Krohn, Mrs. A. C. Lang, Mrs. I. J. Wolf, Mrs. J. Tuteur, Mrs. Ben New, Mrs. William Rothenberg, Mrs. W. Nathan, Mrs. A. Weiner, Mrs. S. Moses, Mrs. S. Reinthal, Mrs. B. Mahler, Mrs. A. Leventhal, Mrs. S. Wohlgemuth, Mrs. H. Strassman, Mrs. Clara Levy, Mrs. A. Palevsky, Mrs. Pollock, Mrs. M. Woldman, Mrs. J. B. Rosenbaum, Mrs. B. L. Gabovitz, Mrs. M. Gordon, Mrs. A. S. Maschke, Mrs. S. H. Volk, Mrs. I. Kessler, Mrs. M. Richman, Mrs. Wallach, Mrs. A Levine, Mrs. Ed McIntyre, Mrs. M. Savlan, Mrs. S. Schultz,

Mrs. M. Eisenstein, Mrs. I. Katzel, Mrs. M. M. Viel, Mrs. Civins, Mrs. M. S. Schussheim, Mrs. A. Wakman, Mrs. I. Volper, Dr. Regina David, Mrs. M. Goldman, Mrs. Edgar Oettinger, Mrs. D. Max.

The Jewish Independent, May 20, pp. 1, 5.

Tuesday, May 24

City's Welcome

Announce Program for Zionist Commission to U.S.

Final preparations for Cleveland's reception of the Zionist Commission to the U.S. were being made Tuesday.

The commission, which is headed by Dr. Chaim Weizman, inventor of TNT, and includes Professor Albert Einstein, creator of the theory of relativity, arrives in Cleveland at 2 p.m. Wednesday. It is to stay two days.

The following program was announced Tuesday by Isaac Carmel, director of the Cleveland Zionist organization.

WEDNESDAY
2:30 p.m. Address of welcome by Mayor FitzGerald at City Hall.
4 p.m.: Parade of 200 autos to Cleveland Hebrew School, 2491 E. 55th st. Welcoming ceremonies by 2000 children.
6 p.m.: Dinner at Independent Order of B'nai B'rith.
8 p.m.: Public meeting at Masonic Hall. Addresses by Dr. Weizman, Professor Einstein and Mayor FitzGerald.

THURSDAY
10 a.m. Conference of Ohio Zionist organizations with the commission at Hotel Hollenden.
Noon: Luncheon in honor of Mrs. Weizmann and Mrs. Einstein given by women members of the Cleveland Zionist organization.

The Cleveland Press, May 24, p. 6.

This was the day that the delegation was supposed to visit Newark, New Jersey. In anticipation, on May 7, Mayor Charles P. Gillen called upon the citizens to decorate their buildings and extend a welcome to the delegation. The program, however, then changed, and the visit to Newark was dropped.

Wednesday, May 25

Einstein Here

Scientist Arrives with Zionist Commission

Cleveland Wednesday saw for the first time Professor Albert Einstein, promulgator of the theory of relativity.

Einstein arrived with the Zionist Commission to the United States, of which he is a member.

Dr. Chaim Weizman, inventor of T.N.T., heads the commission. Its members include both Mrs. Einstein and Mrs. Weizman.

The arrival of the commission in Cleveland is to mark the start of a five-year drive in Cleveland, for $1,000,000 for the Zionist cause. This is to be part of a world drive to raise $125,000,000.

Most Jewish merchants declared a half holiday in honor of the commission, closing their stores at noon.

Met at Station

The commission was met at Union Depot by a reception committee and taken to City Hall, where Mayor FitzGerald delivered an address of welcome.

This was followed by an auto parade to the Cleveland Hebrew School, where the commission was welcomed by 2000 children.

The commission will be given a dinner at the B'nai B'rith at 6 p.m.

A public mass meeting is to be held at Masonic Hall at 8 p.m.

A Typical Professor

Professor Einstein looks the part of the typical professor. He has gray hair which he wears rather long.

He plays the violin. He has his violin with him.

A second prized possession which he brought along is his favorite pipe, a worn briar.

The Cleveland Press, May 26, p. 2.

Einstein is of medium height, with rather broad shoulders. He has a high, broad forehead and large, luminous eyes.

Despite his engrossment in the world of science, Einstein's chief thoughts are with men, and not calculations.

Humanity First

"Science is suffering from the terrible effects of the war," he says. But humanity is suffering worse.

"It is humanity, not science, that should be given first care."

Einstein obtains relaxation from scientific research by reading fiction.

Einstein has one of his chief desires to see an international agreement which will prevent future wars.

Einstein will talk to students at Case School of Applied Science Thursday.

Einstein's theory received its name of the "relativity" theory from his assertion that there is no such thing as absolute time, space or motion. All these, Einstein says, are only relative.

Theory Explained

Developing this Einstein maintains that phenomena such as inertia, gravitation and time are not separate phenomena, but only different aspects of a single unity.

This unity Einstein calls space, and he proceeds to continue time as a fourth dimension in space.

The importance of the Einstein theory lies in the fact that it enables scientists to trace the connection between many phenomena which formerly were inexplainable.

Einstein first began work on his theory in 1905. At that time he was an employee in the Swiss patent office.

He now holds a professorship in the University of Berlin.

The Cleveland Press, May 25, pp. 1, 8.

The "single unity" must of course be "field" and not "space," but let's blame the interpreter. I'm just happy that the German *Feld* was not translated as "meadow."

Cleveland Fights to See Einstein

CLEVELAND, May 25—Only the strenuous efforts of a squad of Jewish war veterans, who fought the people off in their mad attempts to see them, saved Prof Albert Einstein, discoverer of the Einstein theory of relativity, and Dr. Chaim Weizmann, president of the World Zionist Organization, from possible injury upon their arrival here today, city officials, reception committees of Zionist leaders and several thousand citizens thronged Union depot to welcome them.

Most Jewish merchants closed their stores at noon. Prof Einstein and Dr Weizmann, with their wives, were escorted to the City Hall, where in City Council Chamber formal welcoming ceremonies were held. Mayor Fitzgerald spoke for official Cleveland.

Jewish war veterans in uniform, headed by the 3rd Regiment National Guard Band, marched before 200 automobiles.

After the reception the party adjourned to their suite at a hotel. After a brief rest there was another parade, after which they were guests at a banquet.

The Boston Globe, May 26, p. 10.

The *Globe* apparently didn't know that the wives arrived next day (see p. 301).

The Board of Governors of the National Institute of Inventors admitted honorary membership to Einstein as "an inventor of international repute and fame" and sent a letter of the event on this day. To emphasize the prestige of the institute, a few honorary members were also listed: "Major General E. D. Swinton, C.B.D.S.C.H.E, Asst. Secy to British War Minister; Rear-Admiral Fiske, President of Army and Navy League; Lady Duff Gordon, the famous Lucille, the inventor and designer of women's apparel; W. Marconi; Charles Steinmetz; Charles E. Bradley, nitrogen wizard."[86]

Einstein accepted the honor after his return to Berlin. In his letter he remarked that "even though I am aware of the fact that you gave me the honor for my scientific findings, I do not want to leave it unmentioned that I also tried my hand, as a would-be rider, in technology."[87] He must have had in mind his airfoil that was tested even in flight in 1916–17 and flew, according to the test pilot, "as a pregnant duck."[88] His co-honorees look rather militaristic for a pacifist, with the exception of Lucille, whose contributions represented weapons for a battle that Einstein had never been against.

Thursday, May 26

Einstein, in Cleveland, Explains Theory

Professor Albert Einstein, promulgator of the now world famous "Theory of Relativity," explained his theory in Cleveland Thursday. He said:

"My theory changes the fundamental concept that time and space are absolute and infinite.

"It's my belief that if the substance or matter of the universe was suddenly annihilated neither time nor space would exist.

"It is only possible to measure time or space in relation to some other object. For example the motion of a train is measured between stations.

"Put a train out in empty space alone and you couldn't measure its rate of speed or the amount of space it covered.

"All measurements must be in relation to something else. Hence the name, 'Theory of Relativity.'

"Time in my theory is a dimension, I call it the fourth dimension.

"Among the chief ramifications resulting from this viewpoint is that the size of objects varies with their rate of motion.

"Similarly, the passage of time is not absolute but depends upon the rate of motion.

"These differences are microscopic in all ordinary affairs, but in astronomical phenomena where distances and sizes of objects are gigantic they are considerable.

"My theory connects many phenomena formerly regarded as isolated under one head.

"I believe that gravitation and inertia, for example, are space phenomena.

"Gravitation has been thought of in the past as a force. I think of it as a shortening or warping of space into the fourth dimension of course.

"My theory in reality is two theories or better still a double deck theory. I call them the special and the general theory.

"What I have just described is the general theory. This was evolved last—out of a consideration of the facts revealed by the special theory.

"The special theory deals with various astronomical and physical facts.

"One of them was that a ray of light responded to gravity. This was sub-sequently verified by observation of Mars during an eclipse of the sun."

The Cleveland Press, May 26, p. 2.

> Well, the "astronomical facts" are dealt with not by special but general relativity; the light bent by the sun and observed during the 1919 eclipse came not from the planet Mars but from fixed stars. However, never, ever blame Einstein for these blunders!

Zionists Confer

Einstein and Weizman Start $1,000,000 Campaign

Professor Albert Einstein and Dr. Chaim Weizman, head of the Zionist Commission to the United States, were to confer with delegates from all Zionist organizations in Ohio at Hotel Hollenden Thursday.

This is to mark the start of a five-year drive to raise $1,000,000 in Ohio in the next five years as part of a $100,000,000 fund being raised thruout the world to aid the establishment of a Jewish commonwealth in Palestine.

Dr. Weizman Thursday explained the immediate plans of the Zionist organization.

"Our big job now is to aid immigrants to get to Palestine," he said. "There are many Jews, now fugitives from that no-man's-land, Russia, who are eager to go to Palestine.

Urges Development

"Our second job is to make Palestine able to provide these immigrants with the means of support.

"This necessitates development of irrigation projects so that more land can be tilled and the development of more industrial enterprises.

"The school system must also be enlarged."

Einstein and Weizman arrived in Cleveland Wednesday afternoon. A crowd of about 3000 was down at the Union Depot to meet them.

They were taken to the City Hall where Mayor FitzGerald welcomed them in the name of the city.

An address was also made by President C. S. Howe of Case School of Applied Science.

Weizman and Einstein both responded briefly.

The two men spoke again Wednesday night at a mass meeting at Masonic Hall.

Two hundred thousand dollars was pledged to the Zionist cause at this meeting.

Mrs. Einstein and Mrs. Weizman are expected to arrive in Cleveland Thursday.

Plans for entertaining them at a luncheon have been made by women in the Zionist organization. Mrs. Jennie Zwick is in charge of arrangements for the luncheon.

The Cleveland Press, May 26, p. 2.

Zionist Commission to United States Given Enthusiastic Reception in Cleveland

Distinguished Guests Greeted by Mayor Committee of 100 in Council Chamber—Large Throng at Evening Meeting—Addresses by Mayor FitzGerald, Prof. Weizmann and Dr. Howe Mark Reception—Banquet at B'nai B'rith and Mass Meeting at Masonic Hall Features of Program

Cleveland extended an enthusiastic welcome to the Zionist Commission to the United States Wednesday.

An official reception was accorded the distinguished guests in the City Council Chamber in the afternoon, at which were addresses of welcome by Mayor W. S. FitzGerald, Rabbi Benjamin, Dr. Charles S. Howe and others. The mayor's committee of 100 citizens and members of the City Council were present to meet Dr. Chaim Weizmann, president of the World Zionist Organization, and Professor Albert Einstein, world famed author of the theory of relativity, and other members of the Commission.

A banquet was held at the B'nai B'rith, at 6 o'clock in the evening, and following this event the guests were escorted to the mass meeting at the New Masonic Hall, where an immense throng of Clevelanders and visitors from every section of the state gave them a rousing ovation.

The purpose of the tour now being made by the Commission is to obtain support for the Keren Hayesod, the Palestine Foundation Fund, and

Professor Einstein is actively interested in the Hebrew University project. The cornerstone of this building, to be erected on the Mount of Olives, has been laid and funds for its erection are now being gathered.

Mayor FitzGerald was honorary chairman of the city hall reception committee of 100 and William R. Weidenthal, active chairman. Cleveland public officials, rabbis, ministers, scientists and business men were included in the personnel of the committee. The following are members of the general reception and Keren Hayesod Committee: M. D. Schanman, Max Cohen, I. Feigenbaum, M. Speer, Dr. J. Flock, Dr. I. Gittelson, A. Lipkowitz, A. Sachs, M. Slavin, Sol. Hurvits, Jonas Gross, Dr. M. Garber, Aaron Garber, D. Gara, Isaac Carmel, Rabbi Samuel Benjamin, Rabbi Moses Schussheim, Mrs. M. Garber, Morris Berick, S. Rocker, Louis Cohn, Louis Katz, Mrs. Jennie K. Zwick, Lewis Drucker, Samuel Caplan, Leo Weidenthal, S. J. Beck, M. Blittzstein, Juda Feinberg, Dr. H. Tabakin, Mrs. I. Rothstein, Mrs. Charles Ginsberg, Mrs. Max Cohen, M. Herwald, Dr. S. P. Burstein, M. Gandel, Max Efros, I. Kamin, S. Ismach, Joseph Feder, David Idlesberg, W. R. Weidenthal, A. A. Benesch, Dr. I. Zwick, Alfred Steuer, Miss Sarah Allen and Mrs. J. Teaman.

Newton D. Baker is honorary chairman of the Hebrew University Committee. A feature of Professor Einstein's visit was a reception at Case School, where he inspected the laboratory work being done in connection with the theory of relativity.

An event in connection with Cleveland's reception to the Commission was a luncheon at Masonic Hall in honor of Mrs. Weizmann. Mrs. Jennie K. Zwick headed the committee of women that arranged this event as a feature of a jewelry campaign in aid of the Keren Hayesod.

The city hall, B'nai B'rith building and other structures were decorated with the Zionist colors of blue and white and the Zionist flag floated above the Hollenden, headquarters of the guests, with the American flag.

About 3,000 persons thronged the Union Depot at 2 o'clock in the afternoon to greet the visitors on their arrival in the city. From the depot they were escorted to the City Council chamber, which was beautifully decorated with American and Zionist flags for the occasion.

The Council chamber was crowded to capacity and there were cheers accompanied by applause as the guests entered escorted by Jewish War Veterans bearing American and Zionist flags.

As chairman, W. R. Weidenthal opened the meeting and introduced

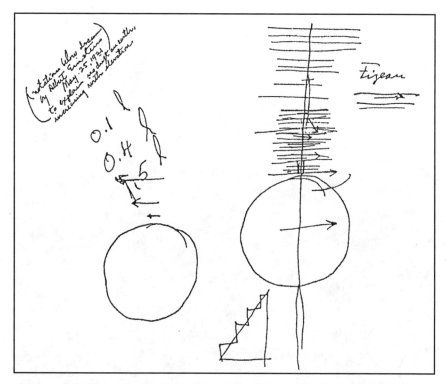

Sketch made by Einstein in Cleveland, Ohio, on May 25 during his discussion with Professor Dayton C. Miller on Miller's Mount Wilson experiments. In the upper left corner Miller remarked: "notation below drawn by Albert Einstein May 25, 1921 to explain no drift on earth increasing with elevation."

Mayor FitzGerald. The mayor declared that he was a believer in Zionism and paid an eloquent tribute to Prof. Einstein and Dr. Weizmann and their work. Among other things the mayor pointed out that where anti-Semitism exists autocracy and oppression are to be found.

Dr. Howe was the next speaker. He spoke of the achievements of Clevelanders in the realm of science and pointed out that Cleveland has always been interested in science. Dr. Howe also paid tribute to the achievements of the visitors. Rabbi Samuel Benjamin was the next speaker and brief addresses by Dr. Weizmann and Prof. Einstein followed.

A parade followed and the visitors were escorted to the Cleveland Hebrew School in E. 55th street. An immense throng awaited the visitors there. B'nai B'rith Hall was crowded to capacity at the banquet which took

Einstein and Ernest Fox Nichols in front of Nela Research Laboratory, Cleveland.
MIT, courtesy AIP Emilio Segrè Visual Archives.

place early in the evening. Intense enthusiasm prevailed and following an address by Dr. Weizmann, who spoke of the need for funds to aid Palestine immigration, an eloquent appeal for pledges and immediate donations was made by Rabbi Benjamin. The response was generous, the total contributions for the gathering being $91,000. From the B'nai B'rith the visitors were escorted to the large mass meeting at Masonic Hall, where the visitors received an ovation. Pledges of about $100,000 followed appeal for contributions to the Keren Hayesod, made at the meeting. Newton D. Baker, former secretary of war, was a speaker at the meeting.

Prof. Einstein remained in Cleveland until 6 p.m. Thursday. He was entertained at dinner by Mr. and Mrs. D. Gara at their residence, 1357 East boulevard, and part of the afternoon was spent by him in visiting Case School. Dr. Weizmann addressed the Regional Conference Thursday.

The committee on Hebrew University met at 10 o'clock Thursday at the Hollenden. Dr. A. S. Maschke is chairman of the committee.

The Jewish Independent, May 27, pp. 1, 7.

Einstein went out to Case to visit the site where the famous Michelson-Morley experiment had been performed. Because Miller knew German rather well, they could discuss the feasibility of repeating the experiment at all seasons of the year. Miller was impressed by Einstein's openmindedness: he was "not at all insistent upon the theory of relativity."[89]

As a photo taken in front of the laboratory indicates, sometime during his stay in Cleveland, Einstein visited Ernest F. Nichols, director of pure science at General Electric's NELA Park Laboratory. They had no problems in communicating with each other in German because Nichols had been assistant to Professor Ernst Pringsheim in Berlin in 1894–96. There he invented the radiometer, which was named after him.

The NELA Park outside the city, named after the National Electric Lamp Association, the previous owner of the site, was considered one of the most remarkable institutions in the field of industrial research; it was often called "the university of light."[90]

19

To Greet Einstein

✦

Tuesday, May 10

Einstein Here Sunday

Plan Official Reception for Scientist and Dr. Weitzman, Zionist

Arrangements for the official reception to Dr. Albert Einstein and Dr. Chaim Weitzman, who will visit this city Sunday and Monday, were made today in Mayor Moore's office.

The Mayor was called upon by a delegation of Jewish citizens, headed by former State Representative Isador Stern. Mr. Moore agreed to preside at a mass meeting to be held in the Metropolitan Opera House Sunday evening, and to appoint a committee of 100 to escort the visitors to Independence Hall on Monday, Memorial Day.

Dr. Einstein, exponent of the theory of relativity, and Dr. Weitzman, head of the international Zionist organization, are in this country to raise funds for [not continued]

The Evening Bulletin (Philadelphia), May 10, p. 1.

Saturday, May 21

Committee Named to Greet Einstein

Prominent Men Will Welcome Author of Theory of Relativity on Arrival in City

Mayor Moore yesterday appointed the Reception Committee to represent the city in connection with the coming to this city on May 29 of Prof.

Chaim Weitzman, of England, president of the Zionists of the World, and Prof. Einstein, author of the theory of relativity.

The committee list follows:

Louis Alexander, Herbert D. Allman, May Aaron, William Abrahams.

Dr. M. Y. Belber, Jacob Billikopf, Gustave L. Blieden, Edward Bok, J. H. Brodsky.

Cyrus H. K. Curtis, Dr. A. J. Cohen.

Morris Daunenbaum.

James S. Elverson.

William Friehofer, Samuel S. Fels, Frederick Felt, James A. Flaherty, Arthur A. Fleisher.

Sigmund Gans, Louis Gerstley, Ellis A. Gimbel, Jacob Ginsburg, Leopold C. Glass, Benjamin Golder, Prof. A. W. Goodspeed, Charles Grakelow, Albert M. Greenfield, Louis Gerson, Joseph Gross, A. B. Goldenburg, J. Gottlab.

Morris Haber, Joseph H. Hagedorn, Max Herzberg.

Alba B. Johnson, J. L. Jones.

Murdock Kendrick, W. Freeland Kendrick, Rabbi Joseph Krauskopf, Solomon C. Kraus, Joseph L. Kun.

Jacob D. Lit, William B. Leaf, Ephraim Lederer, Rabbi B. L. Levinthal, Martin O. Levy, Meyer Lichtenstein, Samuel D. Lit, Harry J. Linsk, Charles Lipschutz, Harry M. Levy, William H. Lewis.

Charles B. McMichael, Benjamin F. Miller, Albert S. Marks, Jules E. Mastbaum, Hiram Maxmin, John Monaghan, William Morris.

John M. Patterson, Josiah H. Penniman, Boies Penrose.

Adolph Rosenblum, Morris Rosenbaum, Benjamin L. Rubinsohn, D. Rosenfeld, B. Rosoff.

Walter Steinman, A. Schley, Harry Sacks, Dr. Jay F. Schamberg, Joseph N. Snellenburg, Maurice J. Speiser, Horace Stern, Isidor Stern, Harry G. Sundheim, David J. Smyth, S. M. Swaab.

David B. Tierkel, H. J. Trainer.

E. A. Van Valkenberg, Samuel M. Vauclain.

Adolph Wachs, Simon Walter, John Wanamaker, David Weber, Richard Weglein, Charles J. Weizz, Samuel Willig, Clarence Wolf, Louis Wolf.

The Public Ledger (Philadelphia), May 21, p. 3.

Philadelphia failed to see Einstein. He decided not to visit there apparently because he was to leave the United States next day, May 30.

According to the *Public Ledger* (May 31, p. 1), when Weizmann arrived on May 29, "more than 15,000 Jewish men, women and children from all parts of Philadelphia . . . jammed Independence Square. . . . With the glaring sunshine beating down upon them and the merciless pushing from all sides, which made breathing almost impossible, between forty and fifty women fainted. . . . The appearance of Dr. Weizmann on the rostrum was the signal for an outburst of cheering and applause such as seldom has been heard in the historic square, which has been the scene of numerous remarkable demonstrations. Men throw their hats in the air and cared little that their headgear was trampled under foot."

I don't think Einstein regretted his absence.

20

Sails Today

✦

May 21

Sometime during the last week of his trip, Einstein sent a letter to his lifelong friend Michele Besso. "America is interesting, with all its hustle and bustle easier to feel enthusiasm toward it than toward countries where otherwise I have a good time. I had to consent to being shown around like a prize ox," he wrote, "to address innumerable small and large gatherings, to deliver innumerable scientific lectures. It's a wonder that I have survived it. But now it's over and what is left of it is the satisfaction that I have really done some good and that I did whatever I could by speaking up bravely and respectfully for the Jewish cause."[91]

May 27

Michael (Mihajlo) Pupin, professor at Columbia University, in his letter of this date apologized for not being able to say goodbye: "Your involvement in the social and political promotion of your talented and much suffered people will serve as a perfect example for other men of science. Our adoration of reason should not withhold us from remembering sometimes that we have a heart, too."[92] Even though he left Serbia early in his childhood, Pupin never forgot to listen to his heart; as chief consul of Serbia in New York, he helped American Serbs to join the Serbian Army in World War I.

His message contained a pack of cigars for Einstein's trip back to Europe.

Monday, May 30

Einstein Sails Today

Dr. Weizmann Will Remain in Interests of Zionism

Professor and Mrs. Albert Einstein will sail for Europe today on the Celtic, leaving behind them some puzzled academic minds. Since he came to this country several weeks ago in the interests of the proposed University of Jerusalem Professor Einstein has been the centre of attraction for scientists who have heard him lecture on his famous theory of relativity. He has spoken at several universities and had the order of Doctor of Science conferred on him by Princeton University.

Dr. Chaim Weizmann of the World Zionist Organization and other members of the commission will remain here for a short time. Mrs. Weizmann, who is President of the Women's International Zionist Organization, which is trying to raise $5,000,000 for welfare work among Jewish women and children in Palestine, appealed yesterday for Jewish women to contribute their jewels and treasure, "gold and silver, new and old," to the fund.

The New York Times, May 30, p. 8.

Einstein Sails for Liverpool

Famous Mathematician Considers "Americans Essentially a Scientific People"

Professor Albert Einstein, noted for his theory of relativity, sailed yesterday for Liverpool on the Celtic, after several weeks in the United States in which he appeared before several universities and worked for the building of the Jewish homeland in Palestine.

On leaving, in a formal statement, he said:

"I would like to say that the respect and admiration I always felt for the American scientists has been greatly increased as a result of my personal contact with them. I have seen sympathetic dealing with the theory of relativity and a truly detached scientific interest in it.

"My work here has been principally confined to college, and, naturally, I am most interested in them," Professor Einstein continued.

"The spirit of intimacy between professor and pupil in American colleges is most interesting. For instance, I saw a student run over to the dean of a university where I was lecturing and slap the old gentleman on the

The Einsteins on board the SS *Celtic* returning to Europe.
Courtesy Corbis.

back. They both seemed delighted at some game that was going on. In Europe a thing like that could cause a sensation. The student would probably be expelled.

He Likes Quiet Place

"I like New York. The Woolworth building, of course, is interesting. You couldn't miss it. But I am fond of quiet little places. Great cities trouble me. I lived in Berlin for a time, but I did my best work in a little Swiss Alpine town.

"I do not care to speak about my work. The sculptor, the artist, the musician, the scientist work because of love for their work. Fame and honor are secondary.

"My work is my life, and when I find the truth I proclaim it. Of course, every new item is met with a certain amount of opposition. They opposed Darwin. They opposed Galileo. They even opposed Newton.

"But my duty as a scientist is to investigate for the truth. When I have found it, I proclaim it. Opposition does not affect my work.

"Americans a Scientific People"

"The Americans are essentially a scientific people. They have intellectual curiosity, imagination and powers of research. Professor Michaelson, of the University of Chicago, is one of the foremost physicists in the world. America must be proud of him."

Mrs. Einstein, in response to a question regarding American women, said:

"I think American women are lovely, of course. There is no distinct type as far as I can see. But they dress remarkably well. All classes of girls dress like European aristocrats.

"For instance, a girl passed me in the street and looked like a queen. She was dressed not only in the best style, but in perfect taste. I thought she was the daughter of a millionaire, but she turned out to be a typist in this hotel. Why, even the working girls dress like countesses. This is a wonderful thing."

Professor Einstein said that he had refused an invitation to be the guest of Lord Haldane while in London, but did not say why.

The New York Call, May 31, p. 2.

> The *New York Evening Post* (May 31, p. 7), the *New York Times* (May 31, p. 14) and the *Jewish Independent* (June 3, p. 1) also published Einstein's praise of American scientists.
>
> The following article is the only one in which Einstein did side in public with Weizmann in his fight with the American Zionist leaders. Through *Der Yidisher Morgen Zhurnal,* a New York newspaper written in the everyday language of Eastern Jews, he spoke his mind to the Jews and not to the wide readership of the "big" press.
>
> Here it is in English translation.

Action of Zionist Leaders Here Is Sabotage Says Prof. Einstein

This Is His Farewell Message to American Zionists through "Morgen Zhurnal"—Is Satisfied with Jews Here and the Outcome of His Visit

Mrs. Einstein Is Homesick of America

Yesterday Prof. Albert Einstein, the great Jewish physicist, Madam Einstein and Madam Vera Weizmann, wife of the Zionist president departed on board the ship "Celtic" for Europe.

Before they left, a representative of the "Morgen Zhurnal" called them, and all three only praised America for the nice reception they were given here, furthermore for the warm response of the American Jews to the Zionist call.

Prof. Einstein also expressed his satisfaction with the results of his efforts here in the interest of the Hebrew University in Jerusalem, in particular with the campaign of American Jewish doctors to raise money necessary for founding a medical faculty at the university. What the doctors here accomplished so far, he declared, guarantees that the medical faculty will be opened.

He also stated that as a result of his visit Jewish committees were organized in other cities as well to promote the cause of the Hebrew University in general, and even American Christians have taken an interest in the campaign.

The American Jewish intelligentsia, he said, surpassed his expectations.

But the great scientist was also pleased by the Jewish masses, in particular by their warm response to the call of the international Zionist leader.

Prof. Einstein expressed also his particular satisfaction with the progress that has taken place in America in the field of science. This progress, he said, is not only the result of America's wealth but also of the true affinity to science that manifests itself here.

The only source of his dissatisfaction is the reaction of American Zionist leaders. "The American leaders," noted the great thinker, "did not play an honest role. They practiced a kind of sabotage."

Prof. Einstein is now bound for England. He will be received there by the universities of London and Manchester and he will deliver lectures. Later he will visit Spain and perhaps China, too. Both countries invited him in the last several years for a visit. His works had already been translated in Chinese in the last several years.

Prof. Einstein did not confirm the last news, only Madam Einstein, without his consent, and she wanted to give information about him. Prof. Einstein, however, came back during the talk and forbade to continue it. "My husband," said Madam Einstein, "treats such news as boastful."

A little bit later, however, he, with a smile, interrupted the conversation of the representative of "Morgen Zhurnal" with Madam Einstein on her American impressions:

"I don't take responsibility for what she is telling to you," he noted.

Madam Einstein talked about America as a lover.

"You overwhelmed us here with your hospitality," she said. She will be very homesick for America. And she would like to be here again. But now she misses her children in Berlin.

The Einsteins have two adult daughters.

Madam Weizmann also praised America, but for her for the second time.

✦

Prof. Einstein delivered lectures at Columbia University, City College, Chicago University, Boston Academy, Princeton University (he was conferred the degree [of doctor] there). And he visited New York University; also Harvard, and the institute of technology in Cleveland.

Der Yidisher Morgen Zhurnal, May 31, p. 1.

> To assess how he was informed by both sides and how he formed his opinion about their dispute would require a complex investigation beyond the scope of this book.
>
> The other members of the delegation would stay for a while, so only Einstein and Elsa left the United States. No crowd bidding them good-bye, no flags. After the tiring month, they didn't need it anyway.

21

Aftermath

✦

June 27

Professor Einstein on His Visit to America

Berlin, F. O. C.

Professor Einstein returned to Berlin last week and addressed a Zionist meeting in the Blüthner Saal on June 27th. The hall was packed and many hundreds were unable to gain admission. Dr. Hantke opened the proceedings and said that it was their desire to begin the work of re-building Palestine on a larger scale than hitherto. What up to till now had been the task of Zionists only, was now to be the duty of *all* the Jews. After Rabbi Dr. Nobel (Frankfort-on-Main) had spoken on the movement from a religious point of view, Professor Einstein, who received an ovation, spoke of the development of the Jewish Nation and commented on his trip to America, which he had undertaken to gain friends and funds for the University in Jerusalem. He said that the reconstruction of Palestine was not merely a philanthropic task for Jews. It represented not merely an opportunity for colonisation, but it was a problem of first-rate importance for the Jewish people. It was not mainly a refuge for Eastern Jews, but the incarnation of the revived national feeling of all Jews. Was it desirable and necessary to strengthen this feeling? In his opinion the reply must be in the affirmative. It was essential that they, above all, should become once more conscious of their existence as a nation; that they should regain the self-respect which was necessary for such a national existence. They must take upon themselves once more cultural tasks which were adapted to strengthen their collective national feeling. It was from this standpoint that he asked them to look at the Zionist Movement. He would like to say a few words about his experiences during his recent visit to America undertaken in the service of the Zionist Movement, in order to promote the Jerusalem University. His greatest experience was that for the first time in his life he had seen the Jewish people. He had hitherto seen very many Jews in large numbers, but he had never seen the Jewish people, neither in Berlin nor elsewhere in Germany. Those Jewish people whom he saw in America consisted mainly of immigrants from Russia, Poland, and Eastern Europe generally. They still possessed a sound national feeling. He had seen how extraordinarily self-sacrificing they were. In a short time they had succeeded in making a medical faculty of the Jerusalem University a certainty. He found that it was generally the middle class, and, above all, the common people, and not those who by their social position or other advantages had risen to the top, who still possessed the sound instinct of their common origin, and who were willing to make sacrifices on a large scale for their people. It was borne in upon him there that if only they were able to secure Palestine as a real centre for the Jewish people, then, despite the fact that the great bulk of the Jewish people will be scattered all over the world, the existence of such a spiritual centre would remove from them that feeling of isolation which weighed upon them. It was this that he regarded as the greatest possible advantage which would result from the reconstruction of Palestine.

Professor Einstein's address was short, but the enthusiasm of his audience was vociferous and could hardly be silenced. Dr. Alfred Apfel, President of the Jewish Juvenile Union spoke as a non-Zionist and said that the Union would help to collect funds for the building up of Palestine. The last speaker was Mr. Kurt Blumenfeld, Secretary of the German Zionist Organisation. He described the present situation in Palestine and begged all to co-operate with the Zionists in their labours.

The Jewish Chronicle, July 8, p. 22.

The *New Palestine* republishes Einstein's speech using the first-person perspective (July 22, p. 8). A short extract is published in the *New York Evening Post,* July 23, p. 3. You can find it in Einstein's *About Zionism* (New York: Macmillan, 1931), pp. 44–49.

July 1

Dr. Einstein Found America Anti-German

Violently So, He Says, Though He Noted That a Reaction Was Setting In

Copyright 1921, by The New York Times Company. Special Cable to The New York Times.

BERLIN, July 1.—Dr. Albert Einstein says he found America violently anti-German, although with evidence of a change of heart taking place, and England eagerly pro-German. These discoveries in relativity he imparted today to a select party at the home of Herr von Winterfeldt, President of the German Red Cross, whose guest of honor Dr. Einstein has been since his return from America and England.

Dr. Einstein told the guests, among whom were President Ebert and the members of his Cabinet, the members of the Prussian Government, the Chief Burgomaster of Berlin and leaders in the business and industrial world.

"In America there undeniably predominated a markedly unfriendly feeling toward everything German. American public opinion was so excited that even the use of German language was suppressed. At present a noticeable change is taking place. I was received heartily by America's learned men and learned corporations. They gladly spoke German and everywhere were mindful with genuine sympathy of the German scientists and institutes with which they maintained so close a friendship before the war.

"In England the impression forced itself upon me that English statesmen and scholars had it in mind again to bring about friendly relations with Germany. The heartiness of the speeches in England could hardly be surpassed. Better times appear to be coming."

The New York Times, July 2, p. 3.

> Well, Einstein was right. Even before he had been invited to join Weizmann, Carl Beck wrote him: "There are in America a large number of serious men thinking well of Germany. There are, however, a large number of men who appear in that light; but in reality are not inclined that way, and the masses [are] fickle and are easily swayed."[93] In his reply, Einstein underlined this information by accepting that "the time is not yet ripe for a scientist living in Germany to visit

America."[94] Note the precise formulation: he is a person "living in Germany." Remember too that Gano Dunn had softened the cold atmosphere around Einstein in Washington by explaining that Einstein was a Swiss and not a German citizen (pp. 136-137). See p. 96 for Slosson's remarks on the wartime opinion in the United States about Germans.

Einstein compares his experience in the United States with that in England. On the way back to Berlin, he stopped there for a few days and delivered lectures.

July 4

A Genius Makes a Mistake

Dr. Einstein, having observed "America" for several weeks, has reached the startling conclusion that there exists here a strong antagonism to "Germany." He admits that he encountered here nothing of the kind himself, and perhaps if he had been in the country longer he would have discovered that the individual German gets in the United States now as always just about the treatment his manners and personal character would earn for him anywhere.

Our feeling for "Germany" is a different matter. By that word we still mean the Government that started and waged a war against the rest of the world, with results which the world knows too well. So far as the new German Government is different from the old one it is viewed by us kindly enough, but we yet have to be convinced that the difference is very great, or that what difference there may be is not in forms rather than in essentials.

Perhaps Dr. Einstein, in spite of his wonderful mind, is as much mistaken, and in about the same way, when he says that "England" is warmly pro-German as he is when he says that "America" is as warmly anti-German. No more than the Americans have the British forgotten the war, its origin or its details. Their grievances against "Germany" are vastly greater than ours, for they were in the war longer and knew it when it was at its worst—when the commission of atrocities was in the execution of a policy deliberately adopted and carried out by what we mean when we say "Germany."

The New York Times, July 4, p. 8.

July 8

Einstein Declares Women Rule Here

Scientist Says He Found American Men the Toy Dogs of the Other Sex—People Colossally Bored—Showed Excessive Enthusiasm over Him for Lack of Other Things, He Thinks

By Cyril Brown

Copyright 1921 by The New York Times Company. Special Cable to The New York Times.

BERLIN, July 7.—Dr. Albert Einstein, the famous scientist made an amazing discovery relative to America on his trip which he recently explained to a sympathetically-looking Hollander as follows:

"The excessive enthusiasm for me in America appears to be typically American. And if I grasp it correctly the reason is that the people in America are so colossally bored, very much more than is the case with us. After all, there is so little for them there!" he exclaimed.

Dr. Einstein said this with vibrant sympathy. He continued:

"New York, Boston, Chicago and other cities have their theatres and concerts, but for the rest? There are cities with 1,000,000 inhabitants, despite which what poverty, intellectual poverty! The people are, therefore, glad when something is given them with which they can play and over which they can enthuse. And that they do, then, with monstrous intensity.

"Above all things are the women who, as a literal fact, dominate the entire life in America. The men take an interest in absolutely nothing at all. They work and work, the like of which I have never seen anywhere yet. For the rest they are the toy dogs of the women, who spend the money in a most unmeasurable, illimitable way and wrap themselves in a fog of extravagance. They do everything which is the vogue and now quite by change they have thrown themselves on the Einstein fashion.

"You ask whether it makes a ludicrous impression on me to observe the excitement of the crowd for my teaching and my theory, of which it, after all, understands nothing? I find it funny and at the same time interesting to observe this game.

"I believe quite positively that it is the mysteriousness of what they cannot conceive which places them under a magic spell. One tells them of something big which will influence all future life, of a theory which only a small group, highly learned, can comprehend, big names are mentioned of men who have made discoveries, of which the crowd grasps nothing. But it impresses them, takes on color and the magic power of mystery, and this one becomes enthusiastic and excited.

"My impressions of scientific life in America? Well, I met with great interest several extraordinarily meritorious professors, like Professor Milliken. I unfortunately missed Professor Michelson in Chicago, but to compare the general scientific life in America with Europe is nonsense."

The New York Times, July 8, p. 9.

> The original interview was published in the *Nieuwe Rotterdamsche Courant* on July 4, but you can read it in volume 9 of *The Collected Papers of Albert Einstein* in English translation. The journalist was the Berlin correspondent of the Dutch newspaper, a young Dutch lady by the name of Nell Boni.
>
> When, on July 7, the German translation of the interview appeared in the morning edition of the *Berliner Tageblatt*, Einstein immediately called the editors, and the next day the newspaper informed its readers that Professor Einstein "absolutely disagrees" with the content of the interview allegedly attributed to him. The distortions in it had reversed the sense of what he had said.
>
> It appears that Einstein did not decide on this revocation on his own, for he was not in a position to learn of the serious consequences of the report so soon. Chances are that he was asked to do so by Erich Marx on behalf of the Bureau of Keren Hayesod for Central Europe in Berlin. On the same day, July 8, Marx wrote a letter to Elsa (!) Einstein, informing her that the Berlin representative of the United Press would call Einstein to discuss an energetic denial of the interview. Marx added that the interview had done serious damage to the Zionist case and caused a scandal in the United States.[95] "Michelson and many others felt hard-hit," Silberstein wrote, expressing the hope that the interview might be an invention of a journalist. To be on the safe side, however, he asked Einstein about this possibility.[96]

July 9

Chicago Women Resent Einstein's Opinion

Men, However, Seem to Agree on "Toy Dogs" and Dominance of Wives

Special to The New York Times.

CHICAGO, July 8.—Professor Einstein's opinion of America, and of American women in particular, as expressed in an interview cabled from Berlin to *The New York Times* yesterday and reprinted in Chicago this morning brought forth indignant protests from Chicago women today. They took particular exception to Professor Einstein's characterization of American men as the "toy dogs" of American women.

Chicago men, however, seemed to agree with Professor Einstein on the dominance of women and the "toy dog" charge, while professors at the University of Chicago contented themselves with a few nervous tut-tuts and the comment that the German scientist had obtained a warped view of America because of the short time he spent here.

"Perfectly ridiculous," said Mrs. Frederick D. Countiss, 1,534 Lake Shore Drive. "The professor must have met a bunch of movie actors when he was over here. If the American men make more money than other men to spend on their wives, that's American luck. I think the right kind of an American is the strongest character in the world."

"I cannot agree with Professor Einstein," said Mrs. Jacob Bauer, 1,100 Lake Shore Drive. "Women have the vote and men are women's equals."

Domination of women over men in America is not a national phase, but a world-wide one, believes William Wrigley Jr. "It's just as true in Germany as it is here," he says. "Einstein is right on the women, and I'm proud of the fact. Aren't the women behind everything man does? Maybe not openly, but they're there just the same. What else can a man do with his money but spend it on his wife?"

Professor W. P. Evans of the chemistry department at Northwestern University said:

"It seems incredible that a man of Dr. Einstein's attainments should make the statements credited to him. If these statements are correct, they go far to prove the fact that, although he understands thoroughly, we hope (don't forget to put in the 'we hope'), the theory of relativity, he has not the essential qualities for judging the scientific and industrial achievements of a great nation."

"Dr. Einstein saw America under very unfavorable conditions," said Professor Millikan of the physics department of the University of Chicago, who was spoken of favorably by Dr. Einstein as a "meritorious" professor, "due to the popular furore over his discovery. I might say that he had a hectic time."

The New York Times, July 9, p. 7.

Probably He Did Say It All

It would be pleasant to believe, and one would prefer to believe, that the reporter of a Dutch paper who asked Dr. Einstein what he thought about America and the Americans misunderstood and misquoted the eminent exponent of relativity. To do so, however, is practically impossible.

On the one hand, neither the Dutch reporter nor his editor had any easily conceivable reason for presenting the doctor's statements incorrectly. On the other hand, it is a well-known fact that high development in one direction usually has undevelopment elsewhere to balance it. Not really surprising, therefore, is it that a great mathematician and physicist should be a superficial and inaccurate observer of ethnological and sociological phenomena.

So there is no need to become excited over the opinions, at once ludicrously and offensively false, that Dr. Einstein expressed in regard to a land and people where he had just been the recipient of a warm welcome and of the high honors his achievements had earned. It is a matter of regret rather than of indignation that he should have been unmannerly enough to sneer at the amount of attention received by him here—that he should have made the mistake of many another traveler less distinguished than he and assumed that he knew America and Americans from a visit to a few of our cities and brief contact with a few persons, most of whom he could not talk to for lack of a common language.

If the doctor had been ignored here, his evident irritation would have been explicable, and severe strictures on our inappreciation would have been deserved. Possible causes of that irritation are the failure of himself and his companion to make more than a partial success of the special mission on which they came to the United States and the antagonism they aroused where they had expected to win full approval and co-operation.

But why expatiate over what, after all is a very trivial matter? What Dr.

Einstein thinks of us should not affect what we think of him in the only way he interests us, and we are not foolish enough to let his criticism of an America he knows little about, and most of that little wrong, make us any the less ready to value his own work at the worth ascribed to it by its few competent judges.

The New York Times, July 9, p. 8.

> The same day, in Berlin, a journalist presented a letter of introduction to Einstein from the editor-in-chief of the *Vossische Zeitung,* a Berlin newspaper. The young journalist was selected because of his excellent qualifications which guaranteed that he could understand Einstein—so the letter said. The editor even promised that Einstein would have the opportunity to check the record of the interview in the afternoon—a promise we all expect from journalists but hardly ever get. The new interview was published the next day. An abbreviated version was published in translation in the *New York Evening Post* on August 1 (see p. 329–331).

July 10

Einstein on Americans

Wherein the Eminent Scientist Failed to Understand Us

To the Editor of *The New York Times.*

The enrollment of students in American universities represents two-tenths of one per cent of the population. The enrollment of students in German universities represents eight-hundredths of one per cent. If Dr. Albert Einstein is correctly quoted in his interview published in *The* [New York] *Times* this morning [July 8], he has challenged the American people to an extremely stimulating survey of its intellectual standards. In seeming derision, he has unwittingly given brilliant endorsement to our experiment in democracy.

Because no European populace would welcome a distinguished scientist with such enthusiasm, Dr. Einstein has completely misinterpreted the popular sensation accompanying his reception here. He could not understand a people mentally alert and youthfully imaginative, thrilling to scientific achievement, recognizing the possible "influence on all future life" of a stupendous new theory, striving to comprehend the incomprehensible, craving more widely than in any other land that knowledge which is the breath and seat of life.

Dr. Einstein is perhaps too familiar with the appalling ignorance of the European masses—that ignorance which so astonished our war correspondents who talked with German prisoners. Hence he charges us with the very "boredom" which might better be applied to our elders abroad.

Grant that we have no few individuals who have attained in pure science the level of some few abroad; forget the names of Edison, Flexner, Bell, Maxim, Sperry, Hammond, Hewitt, Eastman, Westinghouse and a host of others; forget that America has given the world the automobile, airplane, telegraph, telephone, submarine—in fact, that we have brought forward the most nearly conclusive proof for Dr. Einstein's own theory. Take the general intellectual level which Dr. Einstein makes the basis of his comments. And look at the facts.

We have over a million and a half readers of the popular scientific magazines—over three and a half million of engineering and technical magazines are included. What European nation can even approach such figures showing widespread popular interest in science? We have over 50,000,000 readers of the daily and Sunday papers. And where are the European papers that give the same consistent play to scientific developments that ours do?

We had 20,000,000 pupils in the common schools four year ago; a million and a half in the high schools; over a quarter of a million students in colleges and universities. The most quickly available figures show our university enrollment to be .002 of the population, as against .0003 in Great Britain, .0008 in Germany and Italy, .0003 in France. Finally, take the most significant figures of those who are seeking education under difficulties. Three large correspondence schools alone show over 3,000,000 students. One Western university has 20,000 correspondence students. Our evening and business schools claim over 700,000 students; our Summer schools in 1918 had 160,000 enrolled.

No, America doesn't boast a leisure class that dabbles conversationally in science and philosophy. But the figures of reading and educational efforts in this country bear out the belief that Dr. Einstein should have taken the honors bestowed upon him by our public at their face value—as a sincere tribute to his eminence and as evidence of a profound popular interest in the field where he has few peers.

Kenneth W. Payne
Editor, *Popular Science Monthly*
New York, July 8, 1921.

The New York Times, July 10, p. 2.

July 11
A Product of His Education

As Dr. Einstein, himself obviously a survival from an earlier age, and an illustration, in spite of his scientific and philosophic magnitude, of another form of infantilism, chose or was compelled to follow the older order in his criticism, it is little wonder that he is getting scolded just as were Dickens and Mrs. Trollope in their day. More fortunate than they, however, his tirades are received by most Americans, now, as showing the errors of judgment natural in the circumstances to a man of his sort, and particularly to one who, though not German by blood or birth, is a product of training exclusively German.

That fact fully explains the doctor's indignant astonishment at the way women are favored here and his scorn of men who, as it seemed to him, had abandoned, or were abandoning, the divinely instituted position of irresponsible lords and masters over an essentially inferior sex.

Dr. Einstein will not be forgiven, and should not be, for his boorish ridicule of hospitable hosts who honored him because they believed the guarantors of his greatness in his own domain. That he is small out of that domain is a matter of no great consequence, however, for it is a peculiarity shared by most other specialists of like eminence, and in no degree reduces their value to the world.

He has done real harm in only one respect—the next notability of science who comes over here is likely to have a cool and cautious welcome, and so may go home with some real grievances instead of imaginary ones.

The New York Times, July 11, p. 10.

The following article can best be understood as having Einstein's new interview, published in Berlin on July 10, in mind, for it was written after the first news on the interview (the "dispatch") had reached America but before its English version was published on August 1 (see pp. 329-331).

July 12

Explanation Rather Than Denial

Dr. Albert Einstein is quoted in a dispatch to *The World* from Berlin as saying, in regard to the report that he had made harsh statements about Americans and ridiculed the enthusiasm of his welcome here, that "the Amsterdam interview in no way expresses my sentiments. I never made unfavorable comments on the American people and their mode of life. On the contrary, I arrived in Europe overcome by the warmth of my reception in the United States."

Coming, as it does, from a man who is reputable as well as eminent, this must be accepted, of course, at its face value. The Amsterdam *Courant*, however, is a newspaper credited with common honesty. And there is difficulty—in journalistic circles there will be more than difficulty—in believing that its representative could not comprehend what Dr. Einstein said about the Americans. However it may be as to what he says about relativity, and from *The World*'s dispatch it is possible and even easy, to get the impression that Dr. Einstein explained rather than denied the essential parts of the *Courant*'s article.

What he means, perhaps, is not that he did not say what has given offense here, but that it was not meant offensively, and should not have been taken so. That he talked about us to the *Courant*'s representative he admits. Further, he in effect repeats what he was said to have said about his experiences in this country: "What I said about the sensational interest in the theory of relativity, an interest that rests on misunderstanding, applies to every country."

And there is corroboration of another statement in the original article, for in the new one Dr. Einstein says: "If Americans are less scholarly than Germans, they have more enthusiasm and energy, which cause a wider spread of new ideas among the people."

The New York Times, July 12, p. 1, 2.

July 14

On this day Einstein wrote three messages about the ominous interview: two telegrams to the Central Zionist Bureau in London for his past secretary, Solomon Ginzberg,[97] and a longer letter to the editor of *Berliner Tageblatt*.

The telegrams were published on July 22 (see below). Why he sent this information in two telegrams instead of one can only be explained by his anxiety.

In his letter to the editor, which is perhaps only a draft, Einstein provided an explanatory note which he hoped would be published as the final word on the mess. He noted that he had given the ill-fated interview in a personal conversation that the journalist reproduced from her memory. He offered almost no criticism against its content. The German translation, however, with its arbitrary deletions and false emphases gave it a tone contrary to that of the original. As background, Einstein explained that he had expressed his enthusiasm about the intellectual development of the United States and noted in an aside some amusing episodes, "half as jokes," that he had not meant for the public. The "amiable interviewer" included all these episodes in her text, and the *Berliner Tageblatt* picked these remarks and not the passages on Einstein's enthusiasm about America's development.[98]

I don't know whether the letter was sent to the editor; possibly not, because it was not published in the newspaper. The substance of that letter sounds like the true story.

July 18

Professor Einstein Cables Denial of "Interview" Sent from Holland

New York, July 18.—(Jewish Correspondence Bureau)—The Zionist Organization of America is in receipt of a cable from Professor Albert Einstein in which he emphatically disclaims the interview he is alleged to have given a Dutch journalist in which he criticized unfavorably some of the institutions he observed in this country. The text of Professor Einstein's cable as given out by the Zionist Organization follows:

"I disclaim responsibility for unauthorized reports concerning myself and my view with regard to America. I particularly wish to deny the veracity of statements credited to me in one of the Dutch newspapers which seem to be based on a private conversation, horribly distorted.

"I am not disappointed with America nor with the results of the mission of the Zionist delegation, of which Dr. Chaim Weizmann was the head. The tour of Dr. Weizmann in America was a great spiritual and financial success. In general, I may say that my entire impression of America was most favorable."

The Jewish Independent, July 22, p. 7.

July 31

Prohibition Stays, Says Dr. Einstein

Cites It as Example of Success of Serious Effort in American Life

Prohibition has come to the United States to stay, said Professor Albert Einstein in the interview with the Berlin correspondent of the *Nieuwe Rotterdamsche Courant,* excerpts of which were cabled to this country, where they created considerable stir because of the relativity expert's assertion that American men took no interest in anything but work, and for the rest were only the toy dogs of the women.

Professor Einstein's remarks about the scientific life of America, as quoted in the *Berliner Tageblatt* of July 7 from the interview with the Dutch reporter, do not turn out to be as harsh as they appeared in the cabled report. For instance, after saying that he was sorry at having missed meeting Professor Michelson in Chicago, Dr. Einstein continued:

"But to compare the general scientific life of America with that of Europe would be nonsense, just as one cannot at all compare the rest of the life of Europe with that of America. They are just two quite different worlds."

After telling of the deep and favorable impression made upon him by the scientific life of England, which he found more intensive and more pleasant than that of Germany, Professor Einstein spoke with great appreciation of Oxford University and its traditions, and went on to say:

"But I also found Princeton fine. A pipe as yet unsmoked. Young and fresh. Much is to be expected from America's youth. For even though intellectual life by no means plays any rôle as yet, there is, nevertheless, a group among the younger generation that is striving to elevate the spiritual level, and its efforts are sure to succeed. (Here the *Tageblatt* writer inserted a note to the effect that Professor Einstein meant the so-called New Republic Group.) Everything that is taken up seriously in America succeeds. Look at the prohibition of alcohol, for example. It certainly can't be said that the great majority was for the complete doing away with strong drink; there is still a great deal of inveighing against it, but it was put through, and it will stay put, too."

Several days after the publication of the interview in Holland, Germany and the United States, Dr. Einstein gave out a statement to the effect that it had not conveyed a correct impression of his views regarding the American public, and that he had never made any unfavorable comments upon life in the United States, where he had been so warmly welcomed.

The New York Times, July 31, p. 4.

August 1

My Impression of America

by Albert Einstein

The following is a translation of a statement issued by Einstein to the German press on his return home.

The remarks I am reported to have made to the *Nieuwe Rotterdamsche Courant* concerning my sojourn in America are so hopelessly inaccurate that I feel more than justified in giving my own account by way of refuting that of the young Dutch lady from the *Courant* who asked me for a story for her paper.

I lectured at Columbia University, College of the City of New York, and Princeton, and was delighted with the cordiality of the response, not merely from scientists, who could well have been supposed to be interested in my theme, but in professional men of other departments. I must specially emphasize the fact that my auditors found it quite natural that I spoke in German, and, sometimes in a truly touching way, those who conversed with me made every possible effort to use my language. There were hosts of people who referred to their student days in Germany and expressed the hope that the scientific relations with Germany would soon be resumed.

It is thoroughly annoying that these reports should have been spread in regard to what I am alleged to have said about America's interest in relativity. Naturally, the interest of the Americans in relativity is based to a degree on a misunderstanding, but the same applies to Germany and the Germans. There are people, indeed, all over the civilized world who feel that the anti-rationalist tendency of our times finds support in relativity. We have to do here, however, with a rigid and sober theory, an appreciation of which is accessible by no means to a select circle. Relativity is meant for and should be of interest to every individual who can think and who feels inclined to exert himself sufficiently to grasp it.

If it is possibly true that the American public is less recondite than the German, I had occasion to appreciate, on the other hand, the willingness, eagerness, energy, and enthusiasm with which a recognized idea is accepted by great masses of people.

As to Zionism, when I recall the cordiality with which I was greeted by Jew and Christian alike and when I found them all, regardless of creed, willing to hold up my hands in the Zionistic cause, I can only think back on my stay in the United States with a feeling of profound gratitude and regret that we in Europe cannot enjoy the same relations that obtain in that youthful people across the sea.

In the universities I was particularly impressed with the happy relation that exists between the students and professors. I say again that our students may do more individual thinking, but they lack that wholesome spirit so free from skepticism and other sorts of unhealthy "refinement" that characterizes the American student. If our students could only spend more time with the professors and less in reading magazines of their own selection!

Another feature of American life that made a profound impression on me was the utter lack of offensiveness in its manner of giving vent to its patriotism. Politics with the Americans I met seems to be an inner affair; they give it character. Call it "practical" if you please, but the American feels responsible for what is going on in his country. If that is practical, then let us have some of it.

That America is just on the point of assuming the leadership in science is due not merely to the natural wealth and resources of the country but to the additional fact that her men of means are coming more and more to see that it pays to further scientific study in a general way and that they are already indebted to those scientists who have gone before in their own way and on their own resources.

If these simple words of truth will serve to put the false news of the *Courant* at rest I shall be happy. But I make them not for this sake alone. I am trying to give expression to my gratitude and admiration.

The New York Evening Post, August 1, p. 6.

August 5

The Writer of Topics in the New York Times

should in fairness give his sober estimate of the simple, clear statement of Professor Einstein, which appeared in the *New York Evening Post* on Monday. This statement reflects the dignity, the sobriety, the truthfulness of the eminent scientist, and should give a suggestion to the writers who commented on a garbled cablegram of an unauthorized interview as to the proper way to guard the reputation of eminent personages in the public eye. Professor Einstein made the impression upon all who met him of a most considerate, courteous and simply honest man. His scientific achievements were properly acclaimed, and every word he uttered was immediately printed, and made the most of in characteristic American fashion. But after his departure and at the first word that could be seized upon, the same writers and reporters who had acclaimed him sacrificed the personality of the great scientist for a clever phrase or for the writing of a sensational article. True courtesy and hospitality require that American journalism should make right the wrong done to Professor Einstein.

The New Palestine, August 5, p. 1.

August 24

Mr. Einstein: A Correction

Sir:

In an editorial on Carpentier (July 20th) you attribute to Einstein the assertion that he "met one or possibly two persons in the United States who seemed to him to have understood his theory." Permit me to point out that Professor Einstein never said anything of the kind. On the contrary, before leaving this country he publicly expressed his keen appreciation of the intelligent and sympathetic interest that American men of science have shown in the theory of relativity. Considering the fact that for twelve years American mathematicians and physicists in our leading universities have been writing books and scientific articles and giving regular courses on the theory of relativity, the notion that this theory is the esoteric possession of one or two men is too silly to be repeated in a serious journal like the *New Republic.*

Let me also take this occasion to call attention to the fact that there is nothing ingenerous in Einstein's references to American science in the Dutch interview which our newspapers have made a cause celèbre. If Einstein is correctly quoted there, he only said what leaders of the American Association for the Endowment of Science have repeatedly called attention to, viz. that America cannot compare with the leading European nations in the number of scientific men of really first-rate importance. A confirmation of this is to be found in the fact that only two American have won the Nobel prize in science. I am glad, however, to testify from personal knowledge that Professor Einstein has a most generous admiration for these American men of science who, like Michelson, Millikan or Jacques Loeb, are making important contributions to the advancement of human knowledge.
New York City.   Morris R. Cohen.

The New Republic, August 24, p. 356.

> Good old Cohen!
> George Carpentier was a French prize-fighter trying his luck in the United States. What is interesting for us is the first paragraph of the editorial to which Cohen referred.

Carpentier: A Symbol

Is it because big cities favor contagion that a vogue can now become established so readily? We know that when it was proposed to the New York aldermen to give the freedom of the city to Professor Einstein one honest gentleman asked, who is Einstein? But within twenty-four hours even he was saying, Oh, of course, you mean the man who discovered relativity. The word "relativity" might as well have been Skolinkentot so far as the alderman was concerned. Professor Einstein has since observed that he met one, or possibly two, persons in the United States who seemed to him to have understood his theory. But the tag was sufficient. Hundreds of newspaper articles and hundreds of thousands of people said Relativity thinking they meant something. They liked the word. And Einstein had at least enough vogue to be eligible for the freedom of the city of New York.

The New Republic, July 20, p. 206.

> This was a lesson for Einstein on how to treat journalists; that he learned the lesson is demonstrated by the following story.
>
> Ludwik Silberstein received a letter from Einstein, dated August 10, and a short pamphlet attached to it, ready for publication. He found it "so candid" that he read it to his Chicago colleagues and asked them to tell their colleagues and their wives about it.[99] Two of his colleagues even proposed to publish it in *Science*. Silberstein translated the text into English;[100] his translation is closer to the original than a later, well-combed version that was published in one of the collections of Einstein's popular writings, *Ideas and Opinions*. Unfortunately, the manuscript is damaged, and, because in this form it was not published anywhere, the passage in braces below is taken from *Ideas and Opinions*.[101]
>
> Strangely enough, it was not Silberstein but Einstein who gave a title to the text and added the introductory sentence, as if he had been requested by a newspaper or a journal for an article.

August 10 (?)

"Einstein on the Art of Interviewing"

The following humorous passages from a letter of Professor Einstein's to a friend may interest our readers. "If one is publicly called to account of all one has said, whether one has spoken in jest or possibly in momentary irritation, one is often placed in an exceedingly awkward position, though one probably only gets what one has deserved. But the 'awkward position' becomes a well-nigh hopeless dilemma if one is made responsible for words put into one's mouth by another.

"You will ask who is placed in this quandary? Well, anyone who is sufficiently popular or notorious to fall a prey to interviewers. You may smile at this, but as I have experienced it frequently I can explain it to you. Take this case: One day a gentleman of the press comes to you and asks you to give him a few details concerning your friend Mr. N. For a moment you feel indignant at the proposal and you are inclined to refuse, but you know well enough that if you say nothing the disappointed and therefore irate interviewer will probably write thus: 'I visited one of Mr. N's so called best friends to glean some side lights on his character and mode of life, but this personage was singularly reticent. Our readers can draw their own conclusions from this—probably necessary—discretion.' You realize that there is thus no escape and you say: 'My dear friend Mr. N. is a {cheerful straightforward man, much liked by all his friends. He can find a bright side to any situation. His enterprise and industry know no bounds; his job takes up his entire energies. He is devoted to his family and lays everything he possesses at his wife's feet . . . '}

"The reporter makes out of this: 'Mr. N. takes nothing seriously. He has the knack of making people like him, chiefly because he is flattering most people. He is such a slave to his profession that he has no time—doubtless no inclination—for intellectual pursuits. He is weakly indulgent to his wife, in fact he is entirely under her thumb.'

"In the hands of an expert reporter this would be rendered far more drastically, but you can easily imagine that it would annoy Mr. N. intensely when he read it in a newspaper. Realizing that the information came from you he would obviously be sincerely grieved. What would you do in such a predicament? Tell me and I will do likewise."

Notes

✦

1. Hugo Bergmann to Einstein, October 22, 1919, *Collected Papers of Albert Einstein,* English translation (hereafter abbreviated as *CPAE*), vol. 9, doc. 147.

2. Chaim Weizmann to Kurt Blumenfeld, February 16, 1921, Albert Einstein Archives, Hebrew University, Jerusalem (hereafter abbreviated as AEA), 33-345.

3. Kurt Blumenfeld to Chaim Weizmann, February 20, 1921, in Blumenfeld, *Im Kampf um den Zionismus,* 65.

4. Einstein to Paul Ehrenfest, March 24, 1921, AEA 73-256.

5. Weizmann and Tutaev, *The Impossible Takes Longer,* 102.

6. Ibid.

7. Weizmann, *Trial and Error,* 337.

8. Luther P. Eisenhart to Einstein, October 1, 1920, AEA 36-237.

9. Sachar, *A History of Israel,* 98.

10. Polanyi, Keynote Address. For details of the discovery, see also Davidson, *The Sun Chemist,* 30–33, and Weisgal, *Chaim Weizmann,* ch. 4.

11. Popkin, *Open Every Door,* 135–36.

12. Weizmann, *Trial and Error,* 331–32.

13. Fleming, *New York,* 3.

14. For details, see Weizmann, *Trial and Error,* ch. 18.

15. Rosenblatt, *Two Generations of Zionism,* 97.

16. Ibid.

17. Mann, *La Guardia,* 119.

18. "Uproar in the Lecture Hall," *CPAE,* vol. 7, doc. 33.

19. Archive of the Berlin-Brandenburg Academy of Sciences, II-III, Bd. 38, Bl. 55 and Bd. 39, Bl. 18, resp.

20. Weizmann Archives, Rehovot, Israel.

21. Schoener, *New York,* 262.

22. Lipsky, *Memoirs in Profile,* 105.

23. Mentioned in Einstein's reply to Judah L. Magnes, April 18, 1921, AEA 36-841.

24. Reuterdahl, *Scientific Theism versus Materialism,* 289.

25. Stephen S. Wise, "Luncheon with Dr. Weizmann," April 26, 1921, 3–4, S. S. Wise

Collection, American Jewish Historical Society, Brandeis University, Waltham, Massachusetts.

26. Lipsky, *Memoirs in Profile,* 122.

27. "Time, Space, and Gravitation," *CPAE,* vol. 7, doc. 26.

28. Einstein and Leopold Landau to Minister of Education Konrad Haenisch, February 19, 1920, *CPAE,* vol. 9, doc. 317.

29. Einstein to Judah L. Magnes, April 18, 1921, AEA 36-841.

30. Judah L. Magnes to Einstein, April 19, 1921, AEA 36-842.

31. Alfred N. Goldsmith to Einstein, February 21, 1944, AEA 55-144.

32. Cohen, *A Dreamer's Journey,* 186.

33. Morris R. Cohen to Einstein, June 6, 1921, AEA 32-444.

34. Einstein to the Morris Raphael Cohen Student Memorial Fund, November 15, 1949, AEA 32-486.

35. Research Committee of the College of the City of New York to Einstein, April 23 1921, AEA 36-205.

36. W. B. Scott to Einstein, April 4, 1921, AEA 36-231.

37. Oscar S. Straus to Morris Jastrow, April 6, 1921, American Philosophical Society, Philadelphia, Pennsylvania.

38. W. B. Scott to Minis Hays, April 9, 1921, American Philosophical Society, Philadelphia, Pennsylvania.

39. Einstein to W. B. Scott, April 15, 1921, AEA 36-232.

40. Hammond, *Men and Volts,* 373.

41. Alexanderson, "Transatlantic Radio Communication."

42. Charles D. Walcott to Einstein, April 18, 1921, AEA 36-224.

43. Charles D. Walcott to Nobel Committee, "1921 Fysik. Föreslagen Prof. A. Einstein, Berlin," December 17, 1920, Nobelstiftelsen, Stockholm, Sweden, 16.

44. Gano Dunn to Oswald Veblen, May 5, 1921, AEA 36-246. Copy courtesy of Prof. Deane Montgomery.

45. Shapley, *Through Rugged Ways,* 78.

46. See note 44.

47. Louis D. Brandeis to Stephen S. Wise, April 26, 1921, S. S. Wise Collection, American Jewish Historical Society, Brandeis University, Waltham, Massachusetts.

48. Louis D. Brandeis to Einstein, April 29, 1921, AEA 86-014.

49. See Julian W. Mack to Stein, May 15, 1921, Weizmann Archives, Rehovot, Israel.

50. Isaac J. Stander to Chaim Weizmann, April 25, 1920, Weizmann Archives, Rehovot, Israel.

51. Mann, *La Guardia,* 206.

52. Neilson, *My Life in Two Worlds,* vol. 2, 141.

53. Hendrik A. Lorentz to Einstein, March 19, 1921, AEA 16-539.

54. Einstein to Ludwik Silberstein, May 20, 1921, AEA 71-568.

55. Einstein to Hendrik A. Lorentz, June 30, 1921, AEA 16-541.

56. Neilson, *My Life in Two Worlds,* vol. 2, 142.

57. L. B. Sager to Otto Nathan, January 24, 1957, AEA 74-826.

58. Neilson, *My Life in Two Worlds,* vol. 2, 142.

59. Carl Beck to Einstein, December 28, 1920, AEA 43-217.

60. Einstein to Carl Beck, April 8, 1921, AEA 70-632.

61. Edwin B. Frost to Max Epstein, May 20, 1921, AEA 83-528.

62. Paul Warburg to Einstein, April 12, 1921, AEA 36-253.

63. See note 44.

64. John G. Hibben to Einstein, December 24, 1920, AEA 36-242.

65. On this exchange of letters, see *CPAE,* vol. 9, docs. 84, 101, 103, and 229.

66. Memorandum of Agreement between Einstein and Princeton University Press, May 9, 1921, AEA 67-886.

67. Josephson, *Edison,* 445.

68. Conot, *A Streak of Luck,* 429.

69. For a more exhaustive list, see Bryan, *Edison,* 312–16.

70. Theodore Lyman to Einstein April 15 and 29, May 6, 1921, AEA 36-234, 85-222, and 85-220; A. Lawrence Lowell to Einstein, May 11, 1921, AEA 36-222; Einstein to Theodore Lyman, May 4, 1921, AEA 85-221.

71. Frank, *Einstein,* 185–86.

72. Theodore Lyman to Nobel Committee, "1921 Fysik. Föreslagen Prof. A. Einstein, Berlin," October 26, 1920, Nobelstiftelsen, Stockholm, Sweden, 14.

73. Felix Frankfurter to Einstein, May 17, 1921, AEA 36-209.

74. Harvard Student Liberal Club to Einstein, AEA 36-223.

75. George F. Moore to Einstein, June 4, 1921, AEA 35-203.

76. Sarna et al., *The Jews of Boston,* 76.

77. Weizmann and Tutaev, *The Impossible Takes Longer,* 102.

78. Einstein to Jacques Loeb, May 9, 1921, AEA 36-843.

79. Solomon Ginzberg to Stephen S. Wise, May 13, 1921, S. S. Wise Collection, American Jewish Historical Society, Brandeis University, Waltham, Massachusetts.

80. A. Goldstein to Chaim Weizmann, May 20, 1921, Weizmann Archives, Rehovot, Israel.

81. Weaver, *Hartford,* 93, 96.

82. A. T. Hadley to Einstein, February 2, 1921, AEA 36-265; Einstein to A. T. Hadley, February 22, 1921, AEA 36-266.

83. Irving Fisher to Einstein, April 12, 1921, AEA 36-269; Einstein to Irving Fisher, April 11, 1921, AEA 80-439.

84. Bertram B. Boltwood to Ernest Rutherford, July 14, 1921, in Badash, *Rutherford and Boltwood,* 347.

85. Elmer E. Brown to Einstein, April 5, 1921, and May 9, 1921, AEA 84-914, 84-915,

and 84-918; Einstein to Elmer E. Brown, April 23, 1921, AEA 84-916; Slomo Ginzberg to Elmer E. Brown on behalf of Einstein, May 14, 1921, AEA 84-918.

86. Chester E. Clark to Einstein, May 25, 1921, AEA 30-139.

87. Einstein to Chester E. Clark, May 25, 1921, AEA 30-140.

88. Einstein, "Elementary Theory of Water Waves and of Flight," *CPAE,* vol. 6, ed. A. J. Kox, Martin J. Klein, Robert Schulmann, doc. 39, trans. A. Engel.

89. Swenson, *The Ethereal Aether,* 195.

90. Hammond, *Men and Volts,* 344.

91. Einstein to Michele Besso, May 21–30, 1921, AEA 7-335.

92. Michael Pupin to Einstein, May 27, 1921, AEA 19-154.

93. Carl Beck to Einstein, December 28, 1920, AEA 43-217.

94. Einstein to Carl Beck, February 23, 1921, AEA 70-630.

95. Erich Marx to Elsa Einstein, July 8, 1921, AEA 44-389.

96. Ludwik Silberstein to Einstein, July 13, 1921, AEA 21-042.

97. Einstein to Solomon Ginzberg, July 14, 1921, AEA 43-246 and 43-247.

98. Einstein to *Berliner Tageblatt,* July 14, 1921, AEA 43-241. Draft in Ilse Einstein's hand.

99. Ludwik Silberstein to Einstein, September 4, 1921, AEA 21-046.

100. AEA 73-323, 3–4.

101. Einstein, *Ideas and Opinions,* 15–16.

Bibliography

✦

Adler, Cyrus. *Louis Marshall: A Biographical Sketch.* New York: The American Jewish Committee, 1931.

Alexanderson, E. F. "Transatlantic Radio Communication." *General Electric Review,* October 1920: 794–97.

Badash, Lawrence, ed. *Rutherford and Boltwood: Letters on Radioactivity.* New Haven and London: Yale University Press, 1969.

Bernstein, Herman. *Celebrities of Our Time: Interviews.* New York: J. Lawren, 1924.

Blumenfeld, Kurt. *Im Kampf um den Zionismus: Briefe aus fünf Jahrzehnten.* Edited by M. Sambursky and J. Ginat. Stuttgart: Deutsche Verlags-Anstalt, 1976.

Bryan, George S. *Edison: The Man and His Work.* London: Knopf, 1926.

Cohen, Morris Raphael. *A Dreamer's Journey: The Autobiography of Morris Raphael Cohen.* New York: Arno, 1975.

Collected Papers of Albert Einstein. English translation. Vols. 1–10. Princeton: Princeton University Press, 1987–2006.

Conot, Robert. *A Streak of Luck.* New York: Seaview Books, 1979.

Davidson, Lionel. *The Sun Chemist.* New York: Knopf, 1976.

Ebsworth, Walter A. *Archbishop Mannix.* Armadale: Stephenson, 1977.

Einstein, Albert. *About Zionism: Speeches and Letters.* Edited and translated by Leon Simon. New York: Macmillan, 1931.

———. "Message in Honor of Morris Raphael Cohen." *Ideas and Opinions.* New York: Crown, 1982, 79–80.

———. "Interviewers." *Ideas and Opinions.* New York: Crown, 1982, 15–16.

Fleming, Ethel. *New York.* New York: Macmillan, 1929.

Frank, Philipp. *Einstein: His Life and Times.* New York: Knopf, [1953].

Gade, John Allyne. *The Life of Cardinal Mercier.* New York: Scribner's, 1934.

Hammond, John W. *Men and Volts: The Story of General Electric.* Philadelphia: Lippincott, [1941].

Holli, Melvin G., and Peter d'A. Jones, eds. *Biographical Dictionary of American Mayors, 1820–1980: Big City Mayors.* Westport, CT: Greenwood Press, 1981.

Jackson, Frank H. *Monograph of the Boston Opera House.* Boston: Butterfield, 1909.

Josephson, Matthew. *Edison: A Biography.* New York: McGraw-Hill, 1959.

Kessler, Harry. *In the Twenties: The Diaries of Harry Kessler.* New York: Holt, 1971.

King's Handbook of Boston, Cambridge, MA: King, 1878.

Leitch, Alexander. *A Princeton Companion.* Princeton: Princeton University Press, 1978.

Lipsky, Louis. *Memoirs in Profile.* Philadelphia: The Jewish Publication Society of America, 1975.

Livingston-Michelson, D. *The Master of Light.* New York: Scribner's, 1973.

Mann, Arthur. *La Guardia: A Fighter against His Times, 1882–1933.* Philadelphia and New York: Lippincott, 1959.

Mass Charitable Mechanics' Association. *127th Annual Report 1921.* Boston: Barrows, 1922.

Neilson, Francis. *My Life in Two Worlds.* Appleton, WI: Nelson, 1952–53.

Polanyi, John. Excerpt from the Keynote Address to the Canadian Society for the Weizmann Institute of Science, June 2, 1996, Toronto, Canada. *Globe and Mail,* June 10, 1996. Also available at http://www/utoronto.ca/jpolanyi/public_affairs4j.html.

Poor, Charles Lane. *Gravitation versus Relativity: A Non-Technical Explanation of the Fundamental Principles of Gravitational Astronomy and a Critical Examination of the Astronomical Evidence Cited as Proof of the Generalized Theory of Relativity.* New York and London: Putnam's Sons, 1922.

Popkin, Zelda. *Open Every Door.* New York: Dutton, 1956.

Reuterdahl, Arvid. *Scientific Theism versus Materialism: The Space-Time Potential.* New York: Revin-Adair Co., 1920.

Reznikoff, Charles, ed. *Louis Marshall, Champion of Liberty: Selected Papers and Addresses.* Philadelphia: The Jewish Publication Society of America, 1957.

Rosenblatt, Bernard Abraham. *Two Generations of Zionism: Historical Recollections of an American Zionist.* New York: Shengold, 1967.

Sachar, Howard. *A History of Israel: From the Rise of Zionism to Our Time.* 2d. ed. New York: Knopf, 1996.

Sarna, Jonathan D., Ellen Smith, and Scott-Martin Kosofsky, eds. *The Jews of Boston.* New Haven and London: Yale University Press, 2005.

Sayen, Jamie. *Einstein in America: The Scientist's Conscience in the Age of Hitler and Hiroshima.* New York: Crown, 1985.

Schoener, Allon. *New York: An Illustrated History of the People.* New York and London: Norton, 1998.

Shapley, Harlow. *Through Rugged Ways to the Stars.* New York: Scribner's, 1968.

Silverman, Morris. *Hartford Jews, 1659–1970.* Hartford: The Connecticut Historical Society, 1970.

Straus, Oscar Solomon. *Under Four Administrations from Cleveland to Taft: Recollections.* Boston and New York: Houghton, 1922.

Student Liberal Club of Harvard University. *Quinquennial Report, 1921–1926.* Cambridge, MA: 1927.

Swenson, Loyd S. *The Ethereal Aether: A History of the Michelson-Morley-Miller Aether-Drift Experiments, 1880–1930*. Austin: University of Texas Press, 1972.

Weaver, Glenn. *Hartford: An Illustrated History of Connecticut's Capital*. Woodland Hills, CA: Windsor Publications, 1982.

Weisgal, Meyer W., and Joel Carmichael. *Chaim Weizmann: A Biography by Several Hands*. New York: Atheneum, 1963.

Weizmann, Chaim. *Trial and Error: The Autobiography of Chaim Weizmann*. New York: Harper, 1949.

Weizmann, Vera, and D. Tutaev. *The Impossible Takes Longer*. London: Hamilton, 1967.

Index

✦